Charles William Stubbs, Frederick Impey

The Land and the Labourers

Charles William Stubbs, Frederick Impey

The Land and the Labourers

ISBN/EAN: 9783337397036

Printed in Europe, USA, Canada, Australia, Japan

Cover: Foto ©berggeist007 / pixelio.de

More available books at **www.hansebooks.com**

THE LAND AND THE LABOURERS

CONTENTS.

PREFACE TO THE FIRST EDITION.

THE two quotations which I have placed on the back of the title-page of this book—one from an address by the Bishop of Durham to the Co-operative Congress at Newcastle, and the other from a letter by the Prime Minister to Mr. Joseph Arch—sufficiently indicate, I think, the purpose which I have mainly in view in the publication of the following pages.

In the prospect of early Land Legislation, this expression of Mr. Gladstone's opinion with regard to the merits of Small Farming in England is most important, and is indeed likely to meet with frequent quotation, during the discussions of the next Parliamentary sessions, by those Land Reformers, to whom, like myself, a radical revision of the English Land Laws seems mainly desirable in the interest of the labouring population, whose gradual divorce from the soil and consequent pauperisation during the last century and a half has been the parent of some of the most lamentable and mischievous of existing social evils.

It will be evident, however, that the following pages are not intended to support the view of those who anticipate as a result of such legislation the general establishment of a system of Peasant Proprietary in England, much less of those who are now advocating with so much vigour and enthusiasm what is termed

Land Nationalisation, without sufficient care, however, as it appears to me, to define accurately which of the three utterly antagonistic schemes—(1) Compensatory, (2) Confiscatory, or even-(3) Collectivist—they understand by that very high-sounding phrase. A radical revolution in the English Land System I do without doubt most earnestly desire to see; but I trust that it will be a revolution such as that anticipated by Bishop Lightfoot, "beneficent, social, and economic," by which, among other good results, the rural labourer and cottage farmer shall in adopting co-operative institutions be able to secure for himself all the advantages of Peasant Proprietary without any of its corresponding evils.

GRANBOROUGH VICARAGE, BUCKS,
January, 1884.

PREFACE TO THE FOURTH EDITION.

I HAVE added to this Edition, in the form of Notes and Appendices, certain material which seemed necessary to give the book completeness, both as a record of experiments in co-operative agriculture, and also with a view to bring the history of the Allotment movement up to date.

The lately issued Parliamentary Return of Small Holdings in England, which shows a greatly increased acreage devoted to Allotment farming during the last few years, is, as I have pointed out in the text, most encouraging as an evidence, not so much perhaps of the practical value of the Allotment Acts of 1882 and 1887, as of the salutary influence which the public expression of opinion on the subject has exerted upon the Landlord class in persuading them to grant increased facilities for landholding by labouring tenants.

I am still as firmly convinced as I was six years ago—my last two years' pastoral experience in a great Northern town has only deepened my conviction—that the key to the solution of most of the social problems of our great towns in reality lies in the country, that there are few of these problems whose solution is not largely dependent upon such a revision of the English Land System, as shall make it once more economically advisable to increase the amount of English Labour applied to English Land, and concurrently with that to raise permanently "the standard of living" of the English Agricultural Labourer.

Within the last few days a new hope for the future of co-operative industry in the rural districts has been given to social reformers by the vast scheme propounded by General Booth, involving as it does the establishment of large farm communities, and Labour Colonies, as an integral part of his plan for "the salvation of society".

"*Back to the Land!*" is a cry to which, in former days, I have been little disposed to listen. Nearly twenty years close intimacy with the conditions, social and economic, of rural life, have taught me that success in agriculture, even on a small scale, demands qualities of head and hand and heart, which, to say the least, it is quite idle to expect from a merely miscellaneous company of the loafers and slummers and labour-failures of town life. It is too often forgotten by the glib Land Reformers of our city debating clubs, that the efficient agricultural labourer is not in reality the dull chaw-bacon sort of person of a *Punch* cartoon, but one of the most highly skilled of English workmen. To expect, therefore, that the town labourer who has failed can be readily transformed into the rural labourer who will succeed, is to expect miracles.

But, then, General Booth, at least so it is claimed on his behalf, has already worked miracles, and can perhaps do so again. At anyrate, in answer to my current objection, one of his chief advisers writes to me that "the 'General' can guarantee absolute loyalty, implicit obedience, and a wonderful spirit of self-denying love and comradeship among his people". These are new data, I confess, for the solu‧tion of the problem. All the co-operative experiments of which I have given a record in the following pages have failed, when they have failed, owing to deficiency of moral qualities and defects of character.

If General Booth succeeds—as I heartily pray to God he may—in the social revolution which he pro-jects, he will only add force to my long conviction of the truth expressed in these two maxims which I will beg my readers now to ponder, and to read once again as they close the pages of this little book —" The best ultimate success often comes of noble failure. Un-dying hope is the secret of social vision." " Great social transformations never have been, and never will be, other than the application of a religious principle —of a moral development—of a strong and active common faith."

CHARLES W. STUBBS.

WAVERTREE RECTORY, LIVERPOOL,
November, 1890.

PEASANT FARMERS

AND

CO-OPERATIVE AGRICULTURE.

CHAPTER I.

THE LAND AND THE LABOURERS.

" You tell me you have improved the land, but what have
you done with the labourers ? "—SISMONDI.

" Much food is in the tillage of the poor, but there is that
is destroyed for want of judgment."—PROV. xiii. 23.

" We are told that the first effort of the state in agricultural
matters should be to increase the produce of the land. I
firmly deny that. I quite admit that it is an important thing
to increase the produce of the land, but it is not the most
important. It is not so important as maintaining in happiness
and respectability a large class of our fellow subjects."—
LORD SALISBURY: *Speech at Hitchin, Dec. 7th*, 1882.

TWELVE years' work as a country parson in a
Buckinghamshire village have forced upon me
two very definite conclusions. They are these:
I. That of the many urgent social problems
with which at the present moment Englishmen

are confronted, there are few whose solution is not largely dependent upon such a revision of the English Land System, as shall permanently raise the social and economic condition of the English rural labourer.

II. That any permanent elevation of the rural labourer's standard of comfort is impossible, unless there can be effected either (*a*) a great increase in the proportion of small agricultural holdings in England ; or (*b*) the adoption of some system of agriculture, probably co-operative, which shall once more make it economically advisable to increase largely the amount of English labour applied to English land.

It will be my object in the following pages to justify these two conclusions.

In the present chapter, however, I desire to do two things—(*a*) to give a record of certain facts, with regard to the cultivation of small holdings by rural labourers, which have come under my own immediate observation during the last few years ; and (*b*) to call attention to certain deductions which it seems to me may be fairly drawn from these facts.

And, first, as to the facts.

Economic Results of Small Husbandry.

At the close of the year 1873, I divided a portion of my glebe land (22 acres) into half-acre allotments among my labouring parishioners, at an annual rental of 66s. an acre. I have retained two lots, that is to say, an acre of this ground, in my own hands. I have worked it on exactly the same method of husbandry as that of the remaining allotments. That is to say, being heavy clay land, not over well drained, but sloping for the most part to the south and west, the kind of crops we grow are wheat, beans, oats, potatoes, mangold wurzel, carrots, garden vegetables, and so forth. Now, being interested in collecting what facts I could as to the results to be gained from small as opposed to large culture, I have kept accurate accounts during some years of the outgoings and incomings on my one-acre farm, and what has been my result? In the last six years of agricultural depression, my net profit on the acre, after allowing fully for rent and taxes, seed, labour, and manure, has been £3 8s.

Let me give in a tabulated form an abstract of my balance sheets from 1878-1883.

Year.	Outgoings.			Incomings.			Net Profit.		
	£	s.	d.	£	s.	d.	£	s.	d.
1878	10	0	6	16	6	0	6	5	6
1879	13	18	'›	15	1	0	1	2	6
1880	11	18	3	15	6	6	3	8	3
1881	12	7	5	16	1	0	3	13	7
1882	12	3	4	13	15	0	1	11	8
1883	12	13	4	17	4	6	4	11	2
	£71	1	4	£93	14	0	£20	12	8

Average annual capital employed per acre
(outgoings) . . . . . . £12 3s. 6⅔d.
Average Annual Incomings . . . . £15 12s. 4d.
Average net profit or something like 28 per
cent. on the capital invested . . . £3 8s. 9½d.

It may be perhaps useful to give the full
balance sheet for a fairly typical year. I will
take 1881, because in that year one of the
leading agriculturists in my neighbourhood,
Mr. W. Smith of Woolston, did me the honour
to criticise publicly my results, and I can thus
give his criticism and my reply.

Here is the Balance Sheet for 1881 :—

Outgoings.

	£	s.	d.
Wages of Labour	6	7	9
Seed, etc.	1	3	8
Manure	1	10	0
Rent and taxes	3	6	0
	£12	17	5

Incomings.

	£	s.	d.
Wheat, 5 sacks	5	0	0
Potatoes—55 bushels at 2s. . . .	5	10	0
Twelve bushels damaged potatoes, sold			
for pigs	0	6	0
Wurzels, 30 cwt., Carrots, 3 cwt. . .	1	15	0
Beans, 5 bushels	1	0	0
Straw (one ton)	2	10	0
	16	1	0
	12	7	5
Net profit per acre . . . £3	13	7	

A Farmer's Criticism.

And here is Mr. Smith's criticism :—

There are a few items very questionable. Take potatoes, said to have been sold at 2s. per bushel. Why, I bought fine potatoes last autumn at 1s. 6d., and the market has been clogged ever since. Now take straw at £2 10s. Why, I bought a lot last autumn, delivered home at £2. Now look to manure ; bought £1 10s. for beans and roots, including potatoes. Half his land that must get manured to keep on under such a cropping would need 10 tons of London dung yearly, which at 6s. 6d. per ton at the railway station would cost quite 8s. per ton on the land, or £4 for the half-acre. These three items corrected stand thus—

	£	s.	d.
From 55 bush. of potatoes at 6*d*. . .	1	7	6
From 1 ton straw	0	10	0
Extra for manure	2	10	0
Now I must put in interest of money			
on £12	0	12	0
Total	4	19	6
Loss	1	5	11
Mr. Stubbs's balance	3	13	7

Mr. Stubbs must show a better Balance Sheet, or it is of no use for him to come out to show us how to farm. We cannot all buy cheap dung, neither can we all sell dear potatoes or dear straw, and tenants' capital ought to bear interest."

The following was my reply :—

" In the first place let me say at once that of course I am quite prepared to allow that my Balance Sheet would be valueless as an example of average results if it can be proved that I am farming under exceptionally favourable circumstance of purchase or sale. But this is a contention which it seems to me Mr. Smith has singularly failed in proving. He asserts, it is true, that I have sold my potatoes too dear and bought my dung too cheap. But where is the proof of this? I put down my potatoes at 2*s*. a bushel. Mr. Smith says that I ought to have put them down at 1*s*. 6*d*. Why? Because he bought potatoes at that price last autumn. But I was giving the results of *my* farming, not of his. And I contend that 2*s*. was a fair "times price " about here for potatoes of my sort last Michaelmas

I know several labourers at any rate who sold their potato crop from the same field at 2*s.* 6*d.* a bushel, and more than one who got 3*s.* But for my part I do not argue from that that they were exceptionally lucky in their sale. I only suppose that my tenants were a little better tradesmen than their landlord, were more successful, in fact, in acting up to that business maxim of selling in as dear a market as you can and buying in as cheap an one, which I suppose from his argument Mr. Smith does not accept.

"So again as to straw. Mr. Smith says a ton of my straw is not worth £2 10*s.* because he can buy straw, delivered home, at £2.

"In reply I would say that not only did I get that value for it, but I knew that my purchaser was giving the same price elsewhere, and that other of my labouring tenants sold their straw at the same price, to be fetched away, to the regular dealer, and therefore, I suppose, to be sold again at a profit. Moreover one of my tenants, who has only just threshed his corn, has been lucky enough to sell it at £3 10*s.* · And then Mr. Smith must know very well that allotment straw *is* straw, and not straw, couch grass, and thistles.

"Then, again, as to manure. Mr. Smith says *I* want ten loads of London dung at £4 to keep my land in good heart. I reply, why in the world should I buy ten loads of London dung at £4, when I can get ten loads of Gran-borough dung at half that price with as good results?

"By the way, on this question of manure I should much like to ask Mr. Smith, who I am told is a practical man, two practical questions.

"I. If he considers ten tons of London dung the fair allowance per acre for farming land once a year, what does he think of the farming operations of his neighbours hereabouts, whose land, I venture to assert, seldom gets more than five loads to the acre, once in three or even four years? and

" II. What is the special agricultural advantage of the large dung heaps by the roadside, which I so continually see left subject to the action of wind and rain for weeks, with all the best of the ammoniac liquid, assisted by neatly cut channels, draining away for the fertilisation of the weed crop of the nearest ditch?

" I remember to have read somewhere in an Agricultural Gazette that the waste in farmyard manure in England was equal in value to another rental of the land. When I see the wealth of these roadside dung heaps so unintelligently squandered away, I confess I am not surprised to hear it.

" Lastly, as to interest on capital. That may no doubt be a fair point. At any rate I will concede the 12s., or even the 13s. 7d., if he likes, and put my net profit at only £3 per acre, especially if in return he will do me the favour of stating the exact economic reason why tenants' capital should bear interest at five per cent. over and above the fair wages of its superintendence and the net profit on its use."

The criticism of practical men is always valuable, and so I was glad to get Mr. Smith's letter. Whether my reply was satisfactory I must leave my readers to judge for themselves. The following remarks, however, made by two other of my critics last year, were not quite so easy to answer as those of Mr. Smith, but are none the less characteristic. The first remark was made by the son of a farmer (I am glad to think no parishioner of mine) to the man who is good enough to act as bailiff to my one-acre farm—" Why ! what a fool you be, John, to let your parson get anything off his land ! " And the second remark was made to myself by an ingenuous member of the capitalist class—" By Jove, Vicar, what a good plan this of yours is to raise the rents of the land ! "

There is another fact, however, that is worth attention in the above Balance Sheet. It will be observed that the produce of wheat upon the half-acre was 5 sacks, that is to say, at the rate of 40 bushels or 5 quarters to the acre. On the page of the account book from which the Balance Sheet is copied, I observe that I have made the following notes :—

" In this year John Norman grew 9 quarters of oats on his allotment (1 acre) ; and William Tompkins having a dispute with a farmer as to the likelihood of yield of wheat on his allotment, agreed to give the farmer everything over 7 quarters that was threshed out. On measurement at harvest he had to pay the farmer 1 bushel of wheat. In other words, his yield of wheat from 1 acre was 57 bushels." It will be interesting, I think, to compare these figures with one or two well-known standard results. I will give them a tabulated form :—

Produce of wheat per statute acre in bushels.

Farmer's average in Granborough . . 25
English average 26
French average 13
American maximum 19
Mr. Lawes' (high scientific farming) average . 36
Allotment average in Granborough . . 40
Mr. Lawes' maximum 55
Allotment maximum (W. Tompkins) . . 57
English maximum 60

This contrast in point of yield between their own allotments and their masters' fields does not fail, of course, to strike the men. There is a field in this parish which was held some time ago by a farmer at a rental of 11s. an acre. He gave it up because he could do nothing with it. " It was," as he said, "completely wore out." For this field the labourers now give £4 an acre. But they think they do badly if they do not get the allotment average given above, of 40 bushels of wheat to the acre, while the farmer is satisfied with 25 bushels at most. When one remembers, too, that in very many cases the rent given by the

farmer is little more than half that given by the
labourer, can we be surprised that the labourer
and the farmer differ very widely as to the true
cause of the agricultural depression of the last
few years? And that leads me to the first
deduction which I venture to draw from the
foregoing facts. It is this. That one chief
and much overlooked element in the agricultural
depression of the last few years has been *the
labour-starving of the land* on the part of the
farmers.

The sequence of events in this neighbourhood, at any rate, is not difficult to trace. The
agricultural labourers' agitation of ten years ago,
under the leadership of Mr. Arch, succeeded
in raising the wages from 12s. to 15s. a week.
The farmers protested that they could not afford
to pay the extra wage. They were not able,
however, to resist the pressure of the Union,
but were compelled to give the extra 3s. a week.
But they avenged themselves, as they thought,
by employing less men. At first the plan seemed
every way excellent. A farmer employing
10 men, knocked off 4, and thus saved £2 8s.
per week on his labour bill. To the remaining
6 men he gave the extra wage of 3s., or an

increase of 18s. on his weekly labour bill. The net gain per week to the farmer *in money* was thus 30s., and his net loss *in men* was 3 labourers. *But the money was in his pocket, and the men were out of sight.* This was all very well for the farmer ; but how about the land ? "Ay, there's the rub!" It was starved for lack of labour. Then came the wet years, when more than ever labour was needed. But the labour was not now to be had. It had been driven out of the country. The census of 1881 showed a large decrease in the population of all the villages in Buckinghamshire. The land got fouler and fouler, and the natural result followed. I know of a field of 4 acres on one farm in this parish, which in old days used to bear fair crops, which gave a total yield last year of 22 bushels of wheat, or a little over 5 bushels to the acre. During the whole of these years, however, under exactly similar conditions of weather, the labourers' allotments prospered, with a net result, as we have seen, in one case at any rate, of nearly 28 per cent. on the capital invested.

So much for my first deduction. But there are others. For example, with regard to the

advantages of small holdings, to my mind
three things at any rate are clear.

First, that the possession of a small holding
of land adds very largely indeed to the annual
income of the rural labourer.

Secondly, that small proprietorship, or even
tenancy of the soil, exercises a very beneficial
influence upon the moral character of the agricultural labourer.

And, thirdly, that the system of large allotments or small holdings is worthy of extension
for national reasons, as tending to restore that
lost *balance of property* in the soil which is so
necessary a factor in the civil policy of any
soundly constituted state. Let me say a word
or two on each of these points.

Economic Results of Small Husbandry.

That the occupation of an allotment of land,
if only say half an acre in extent, adds very
materially to the annual income of the rural
labourer, the figures which I have given above
I hope conclusively prove. Incidentally, moreover, this fact suggests a solution to one of the
most difficult problems which the rural reformer
has to face. I mean the Cottage Question.

Labourers' Cottages.

When Sir James Chettam in George Eliot's
" Middlemarch " takes objection to Miss
Brooke's generous schemes for cottage building
on the ground that labourers can never afford
to pay a rent sufficient to return a fair interest
on the capital invested, and yet amiably allows
that "perhaps after all the work may be worth
doing," we are, most of us, I suppose, ready to
give some sympathy to the impetuous outburst
of Dorothea's indignation when she cried—
" Worth doing ! Yes, indeed, I think that we
deserve to be beaten out of our beautiful houses
with a scourge of small cords—all of us who let
tenants live in such sties as we see round us."

But after all Sir James Chettam is only
typical of his class. " That good cottages
cannot be built to pay," is the common burden
of all the squires. The echo of it we have all
heard in Lord Salisbury's recent article in
The National Review on " The Housing of
Labourers," where he states that two-thirds
only of the cost of cottage building can be
regarded as a commercial investment, the
remaining third must be regarded as a charity

and benevolence on the part of the landlord.
And the squires, I believe, are in the main
right. Good cottages, apart from charity, are
impossible for the labourer without increased
wages. But there is one consideration on this
point I would venture to make.

Might not the landlord himself do something
towards increasing the wages by insisting upon
the higher rent? As the farmer has already
found the increased labour bill of the last few
years a severe strain upon his capital, a still
further increase would only be possible by a
corresponding reduction in farm rent. As,
however, the landowner in this case would
recoup himself for the loss in farm rent by a
corresponding gain in cottage rent, it does seem
a little strange that so obvious an economic
adjustment should not long ere this have been
made, at least by those landlords who have so
far recognized the responsibilities of property
as to provide efficient cottage accommodation
on their estates. Their omission to do so I
suppose is in reality a part of that reluctance,
still so common in the rural districts, to take
any step which would seem to substitute in
rural class relationships the principles of com-

mercial justice for those of feudal beneficence. Is it not time, however, that some protest should be made against an economic arrangement which in effect compels the farmer to pay to the landlord what is in reality due to the labourer, in order not to interfere, on the one hand, with that luxury of paternal protection which an improving landlord at present gains by being enabled to grant at somebody else's expense to the agricultural labourers on his estate efficient cottages below cost price, and, on the other, with the argument, so convenient to those landlords who have no desire to house their labourers well, that " cottage-building does not pay " ?

There remains, however, the method, suggested by the results attainable on successful allotment farming given above, by which, as it appears to me at any rate, cottage property even at the present rate of wages can be made to return a fair interest on the capital invested.

Let the landlord grant to every cottage tenant an allotment of land of not less than one acre. A cottager who cannot afford to pay 2*s.* a week for a cottage only, can well afford to pay 3*s.* 6*d.* or 4*s.*, or even more, for the same cottage with an acre of land attached. It is my contention, therefore, as the second deduction I would desire to draw from facts, that although a cottage without land cannot, under present conditions, be built to pay, a cottage with land can.

Moral Results of Small Holdings.

The cottage question leads, of course, directly to moral considerations.

For how, I would ask (and I think here I have a right to put the question from my experience as a country parson),—how under such physical conditions as those which I find surrounding me in my daily work among my labouring parishioners—where, to put the matter as briefly as possible, in a village of fifty cottages we have only three with more than two bedrooms, and seventeen with only one—is it possible for me to expect from my parishioners any approach to that "pure religion breathing household laws," which it is yet my duty as their pastor to endeavour to inculcate ? How, with mere huts for homes, can the distinctively home virtues find any room for growth, parental love, filial obedience, household thrift, cleanliness, modesty, chastity, self-respect, purity, and simplicity of heart? Can I honestly ascribe the meagre growth of these virtues among my people solely to failure of individual will, or must I not rather trace them to circumstances of life and sleep so degrading as to leave no moral room for their growth? What provision can there be under such conditions of home life, not only for the three essentials of physical life, pure air, pure water, pure food, but also for the three essentials of spiritual life, " admiration, hope, and love "?

But again, look at the moral results of the occupation of land by the rural labourer from another point of view, namely, its influence on the promotion of thrift. And here I would desire to lay stress upon a principle which it appears to me is too much overlooked by those who are always preaching thrift to working people. It is this,—There is a stronger motive to save created by the desire of investment in the present than by the desire of insurance against the future. In other words, we shall be more sure of success in any attempt to encourage thrift and frugality among labouring people, if we can show them any means of lucrative investment, open to those of them who will exercise self-denial and economy, than if we were merely to point out to them the advantages of insuring against the probable misfortunes of the future.

It is not of course, I need hardly say, that I would discourage such efforts at insurance—far from it. Penny banks, medical and sick clubs, provident dispensaries, clothing clubs, life insurance societies—all these of course it is our duty, by all the means in our power, to encourage. These are indeed the "working plant," so to speak, of every well-ordered parish.

But I still desire to press home this principle, that we cannot afford to neglect that greatest of incentives to thrift which is created by the opportunity of direct present investment of savings.

And that opportunity may, I am sure, best be found for the rural labourer in the occupation and cultivation of land.

I know nothing, at any rate, which fires the imagination of the rural labourer more than does that opportunity. To my mind, it is the natural starting-point in any successful scheme for the depauperisation of the labouring population.

After all, remember land is the most natural savings bank of the agricultural labourer. It is a bank which he understands. He is familiar with its working. He knows something of its system of deposits, its method of exchange, the nature of its reserve fund, of its risks, of the rates of interest which it offers, the value of its securities.

Mr. Cobden, in a letter republished by Mr. Bright not long ago in the *Times*, in fact sums up this side of the question in these words : " Looking at the moral aspect of the question alone, no one will deny the advantages which the possession of landed property must confer upon a man or a body of men—that it imparts a higher sense of independence and security, greater self-respect, and supplies stronger motives for industry, frugality, and forethought, than any other kind of property." Mr. Cobden's remarks refer, of course, especially to the continental system of peasant proprietorship ; but they apply also, though not perhaps to an equal extent, to the English system of allotment tenancy. In the first report of the Royal Commissioners on Agricultural Employment in 1867, ample evidence will be found in confirmation of the statement that the occupation of land influences most beneficially the moral character of the labourers.

Social Results of Small Holdings.

But there is also a most important social side to this question. At present the allotment system is in fact the only means that has as yet been adopted to check that gradual alienation of the rural labourers from the soil of England which has been going on during the last century and a half, with, I venture to say, such mischievous social results. There is no man, I am sure, who has any care for the past or the future of his country, who can read any of the summaries of the Domesday Books, published a few years back, without mourning over the decadence and the national extinction of the independent class of small holders. Our tenant farmers, whatever their industrial value, do but ill-bridge over the social chasm between the landowner and the labourer. " What is to be the future land system of England ? " is the question that in reality faces us. How shall we answer it ?

We stand, in fact, at the parting of the

ways. On the one side we have the advocates
of Individualism, on the other we have the advocates of Nationalism. In New Zealand, I think
I should be a Nationalist, for there the nationalisation of the land might be possible as well as
wise. In England the scheme, as it appears
to me, would be neither wise nor possible
short at least of civil war, and here therefore
I can be a supporter of neither the drastic
proposals of Mr. Henry George, nor the milder
plans of Mr. Russell Wallace. But I can be
an Individualist none the more. In economic
questions I must still give the first place to
moral considerations. In the eyes of God, I
cannot forget that society exists not merely to
further the accumulation of capital, but for the
sake of the well-being and the happiness of the
individuals who compose it. In considering,
therefore, the possible future of the English
land system, it is not only of the produce of
the land or the profits of landowners that I am
thinking. The well-being of the people is not
of less importance than the wealth of the col-
lective body. By the system of adding field to
field, much has no doubt been gained by the
state. But has nothing been lost ? The gain

may be measured by roods and perches, by pounds and by shillings and by pence. But with what measure shall we measure the loss? "You tell me you have improved the land," exclaimed Sismondi, "but what have you done with the labourers?" In the last resort, after all, the question is not about wealth but about men. Or if we must think of things from the selfish point of view, consider this question. In a short time we shall have enfranchised a million of farm labourers. How are they to be secured on the side of public order?

A wide extension of proprietorship in the soil is, I answer, the strongest bulwark of national safety. Those who talk about the danger of Radical and Socialistic ideas, appear to forget, that when a social commune was erected in Paris in 1871, there were five million land-owners in France ready to take the side of public order, and to enforce the conservative view with regard to the right of property. Have we any such conservative safeguard in England? I venture to say that the seven hundred and ten landowners who are proved by the new Domesday Book to hold one-quarter of the whole land of England, could not stand for one moment against the breath of revolution. I have no desire to be an alarmist. But I do most solemnly believe that the concentration of land in large estates among a small number of families, and the consequent accentuation of the contrast between the rich and the poor, is full of danger for the future, and is, in fact, a direct provocative of social revolution. *Latifundia perdidere Italiam*, was the verdict of the historian Pliny on the ancient Empire of Rome. God grant that it may not be the verdict of some future historian on the fall of Imperial England !

NOTE TO CHAPTER I.

Results of Allotment Farming.

I think it will be instructive to place beside the results of Allotment Farming in Bucks, as exhibited in the Balance Sheets of my Granborough Allotments for the years 1878-1883, the following results which have been kindly sent to me by Mr. Tuckwell, the Rector of Stockton near Rugby, under date August 24, 1890.

SOME BALANCE SHEETS FROM THE
STOCKTON ALLOTMENTS.

" WITH all the current talk about Allotments, few persons understand the capacity of the land for food-producing and money-saving when cultivated by Labourers for themselves on fair economic conditions. My own platform assertion that two acres of Allotment, wisely handled, mean a certain saving of

SEVEN SHILLINGS A WEEK

throughout the year, is met with angry contradiction from Conservatives who are opposed to the Labourers' emancipation, and from Farmers whose conception of land value is vitiated by generations of slovenly farming; while it finds doubtful adhesion from Radicals who fear that my estimates are speculative and romantic, and from Labourers whose experience has been confined to a few poles of garden ground.

" I am asked to set down actual returns of

ANNUAL EXPENDITURE AND INCOME

from our Stockton Allotment holders during the last two years. I take four specimens; they are fairly representative, and the individual cases can be verified by visitors to the farm. It will be useful to bear in mind that our average crop of wheat is 12 to 14 bags per acre (the bag being 3 bushels), of beans about the same, of potatoes about 200 pots or bushels to the acre.

" Most of our Allotment Tenants grow

FOR CONSUMPTION, NOT FOR SALE;

but I give instances of both ; estimating the selling price of wheat at 10s. to 12s. a bag, of beans at 10s. to 11s. a bag, of wheat straw at 1s. 6d. to 2s. 3d. a cwt., of potatoes at 2s. a bushel. To sow an acre, 2 to 3 bushels of wheat, 4 bushels of beans, are required ; thrashing costs 1s. 6d. a bag. Note also, as important factors in our calculation, that a bag of wheat yields when ground 35 loaves worth 4½d. each, that the "offal" of the wheat pays for the grinding, that a freshly killed pig sells for 9s. a score, but that bacon cured and eaten represents 8d. per lb., or 13s. 4d. a score; that one pig gives at least 3 loads of manure in the year, and that 12 to 16 loads of manure are applied to an acre in two years.

Now for my Instances.

" Joseph Gardiner holds half an acre. It gave him 5 bushels of potatoes, and entirely fed a pig, which when killed weighed 18 score and was sold.

	£	s.	d.		£	s.	d.
Rent	0	10	0	Pig sold at 9s. a score	8	2	0
Rates . . .	0	1	9	Potatoes . . .	0	10	0
Ploughing . .	0	9	0				
Pig	1	0	0				
Seed . . .	0	2	6				
	2	3	3				
Balance . . .	6	8	9				
	£8	12	0		£8	12	0

" His net profit was £6 8s. 9d. in the year, 2s. 5½d. per week, on half an acre.

" Edward Smith holds 1 acre, which yielded 44 bushels of wheat.

	£	s.	d.		£	s.	d.
Rent	1	0	0	3 bushels sold . .	0	11	0
Rates . . .	0	2	6	41 bushels eaten .	8	10	7
Ploughing . .	0	18	0	Straw sold . .	3	0	0
Thrashing . .	1	0	6				
Seeds . . .	0	7	2				
Balance . .	8	13	5				
	£12	1	7		£12	1	7

" His net profit on 1 acre was £8 13s. 5d. in the year, 3s. 4d. a week.

" Richard Shelton holds 1½ acres. He grew 5 tons of clover.

	£	s.	d.		£	s.	d.
Rent . . .	1	10	0	Clover—5 tons at £4	20	0	0
Rates . . .	0	4	3				
Seeds . . .	0	7	6				
Getting in crops .	3	0	0				
Balance . .	14	18	3				
	£20	0	0		£20	0	0

" His net profit on 1½ acres was £14 18s. 3d. in the year, 5s. 8¾d. per week.

" Job Wincote holds 2 acres.   3 bags of wheat, 9 bags of beans, went to feed the pigs, and are omitted.

	£	s.	d.			£	s.	d.
Rent. . . .	2	0	0	6 bags of wheat sold				
Rates . .	0	5	0	at 12s. . .		3	12	0
Ploughing. .	1	16	0	3 bags of beans sold				
Thrashing .	0	18	0	at 12s. . .		1	16	0
Seed. . .	1	3	6	1 bag of green garden				
Pig . . .	2	10	0	peas . . .		1	4	0
Balance . .	31	9	4	15 bushels of potatoes		1	10	0
				15 cwt.of wheat straw,				
				sold . . .		1	2	6
				15 ditto kept for				
				thatching . .		1	0	0
				2 pigs (bacon) 35 score				
				16 lbs. . .		23	17	4
				6 young pigs, sold .		6	0	0
	£40	1	10			£40	1	10

" His net profit on 2 acres was £31 9s. 4d. in the year, 12s. per week.

"I have many more accounts before me, but these are enough to show what can be done, what *has been done*, by Labourers working in their spare time upon their own land. The figures will probably seem startling. The four cases summarized show a *net profit per acre of £12 5s. 11½d. in the year, or 4s. 8¾d. weekly throughout the year.* The average profits of the entire Farm would at present certainly not amount to this, but I believe that they will do so as our Tenants become each year better educated by Allotment experience.

" It will be objected that in these Tables I made no allowance for

LABOUR WAGE.

In estimating *profit*, certainly not, for the essence of Allotment success is that the labour costs no money, being the product of spare time which would otherwise be wasted. In estimating *rent* on the other hand, labour wage must be admitted, rent being what the land can spare to the landlord after a fair profit has been obtained by the cultivation. Already it has been said to me by landlords who have seen my figures—" If my Allotment holders can secure so large a profit, it is fair that I should raise their rent ". The following Table shows the cost of time and labour spent on 1 acre in a year as the rate of ordinary farmer's wages, 3d. an hour.

		£	s.	d.
Wheeling and spreading 8 loads of manure				
at 1s. 6d. a load . . . . . 48 hours .		0	12	0
Digging at 48s. per acre . . . . 180 ,, .		2	8	0
*Sowing at 10s. per acre . . . . 36 ,, .		0	10	0
Three Hoeings, cheap at 15s. per acre . 72 ,, .		0	15	0
Fagging at 12s. per acre . . . 24 ,, .		0	12	0
Loading and carrying 7s. per acre . 12 ,, .		0	7	0
Total in one year . 372 ,,		£5	4	0

[* *Pegging* would be 13s. an acre ; *drilling* would be 7s. 6d. I have struck a medium.]

" The next Table will show the fair commercial profit on an acre, *calculated as a farmer would calculate it*, and calculated for obvious reasons upon two years' work.

	£	s.	d.		£	s.	d.
Seed wheat,2½ bushels	0	10	0	12 bags of wheat at			
Seed beans, 4 bushels	0	16	0	11s. . . .	6	12	0
Thrashing wheat,				10 bags of beans at			
1s. 6d. per bag .	0	18	0	11s. . . .	5	10	0
Thrashing beans,				Wheat straw, 29 cwt.			
1s. 3d. per bag .	0	12	6	at 1s. 9d. . .	2	10	0
Manure, 16 tons .	4	16	0	Bean straw, 25 cwt.			
Profit without rent				at 1s. 3d. . .	1	10	0
or labour wage							
on two years .	8	9	6				
	£16	2	0		£16	2	0

"Now, put these tables together. The net profit without rent or labour wage is £4 4s. 9d. per acre per annum ; it is more than swallowed up by labour wages. leaving less than nothing for rent. This is the *commercial profit*; by dexterous management and by consumption instead of sale the Allotment holder doubles it ; but the *actual profit* thus obtained is due to personal sacrifice and cleverness, which the landlord has no right to confiscate.

"And let it be remembered, lastly, that the present success of these Tenants is due to righteous conditions of tenure, to the land being conveniently situated, held at fair rent on a long lease, and in such quantities as each labourer requires. We have probably in England a million agricultural labourers longing for land upon these terms. We have land everywhere available for them. How long will the unjust laws survive in the interest of seignorial selfishness and class pride, which have driven our villagers from their heritage in the past, and seek still to perpetuate their exile ? "

CHAPTER II.

" Se l'aratro ha il vomero d'argento, la vanga ha la punta d'oro " (*Though the plough has a silver share, the spade has a golden edge*).—*Italian Proverb.*

" As long as the connexion of the peasantry with the land was unbroken, England was perfectly free from every symptom of pauperism."—W. T. THORNTON.

" The question in the last resort is not about wealth, but about men."—ADOLF SAMTER : *Social-Lehre.*

As far back as the first year of this century a gold medal was offered by the Board of Agriculture to the person " who shall explain in the most satisfactory manner the best means of rendering the allotment system as general throughout the kingdom as circumstances will admit," which the Board asserts to be a " great national object." During the early years of the century " The Society for Bettering the Condition of the Poor "—among the earliest members of which was Mr. Wilberforce—urged, in their annual reports, the advantage of attaching suitable allotments to cottages, as one of the means by which the agricultural labourers would be improved both physically and morally. In the years immediately succeeding the peace of 1815, when the necessity for Poor Law Reform became so evident, the allotment system became a prominent subject with all social reformers. Mr. Cobbett, in his " Rural Rides," notices, in the year 1821, the benefit to the labourers of the good gardens which he observed in several parts of England, especially in the southern counties. " There," he says, " you see that most interesting of all objects, that which is such an honour to England, and that which distinguishes it from all the rest of the world, namely those neatly kept and productive little gardens round the labourers' homes, which are seldom unornamented with more or less of flowers. . We have only to look at these gardens to know what sort of people English labourers are." In 1827, a witness, examined by a select committee of the House of Commons on emigration, stated : " I could load the committee with information as to the importance of the cottagers renting a portion of land with their cottages ; *it keeps them buoyant, and it keeps them industrious.*" And he enforces his opinion of the duty of placing such land freely within their reach, on the ground, that "since 1760 they had lost about 4,000,000 acres of common, which they had formerly the privilege of using for their pigs, geese, and a variety of other things."

Action of Legislature.

In 1819 an Act of Parliament was passed empowering the churchwardens and overseers of any parish to purchase or take on lease any suitable portion of land, and to let such portion of land " to any poor and industrious inhabitant of the parish," to be occupied and cultivated on his own account. In 1831-32, further Acts of Parliament were passed, either amending or extending the provisions of the Act of 1815.

The effect of these Acts of Parliament, and the strong opinion expressed by the Poor Law Commissioners in 1834, " that the immediate advantage of allotments is so great that if there were no other mode of supplying them, we think it would be worth while, as a temporary measure, to propose some general plan for providing them," was to extend greatly the allotment system throughout the country. A select committee of the House of Commons sitting in

1843, reported that "the tenancy of land under the garden allotment system is a powerful means of bettering the condition of those classes who depend for their livelihood upon their manual labour," and they added the important opinion that "its benefits are not obtained at the expense of any other class, nor accompanied by any corresponding disadvantage." Two years after this very favourable report of the select committee, Mr. Cowper, in introducing a Bill to 'Promote the letting of field gardens to the labouring poor," stated that, " It appeared from history that before the land of England was brought fully into cultivation, almost all cottagers had land for tillage. All those above the condition of serfs had land in their own occupation, and, in addition to that, had common right over the waste lands. . . . He believed that previous to the 16th century all the peasantry drew portions of their maintenance from the soil. Since 1800, no fewer than 2,000 Inclosure Acts had passed. The amount of acreage was not set forth in the returns, but it must form no inconsiderable portion of the land of the country. The consolidation of small farms, so extensively adopted during the war with France, had con- .

tributed to deprive the labouring man of his
opportunities of holding land. The giving up
the tenure of leases on lives also had the same
tendency. The result of the combined causes
was that, until the allotment system was revived,
the English labourer was severed from all connexion with the land. . . . What he particularly
valued in the system of allotments was the
moral effect on the holder. The management
of a garden was an important ingredient in his
happiness. It was just the amusement which
suited the labourer, and for which he was suited.
This amusement was elevating in its tendencies;
and many idle, careless, lawless individuals
would be converted into steady, sober, industrious men, by having the means of harmless,
rational, and profitable employment."

This Bill, however, though it succeeded in
passing the House of Commons, and was read a
second time in the House of Lords, was finally
allowed to drop, Mr. Cowper expressing it as
his opinion that notwithstanding all that had
been done, he believed a generation might be
expected to pass away before there would be
a general allotment of garden ground for
labourers.

Considerably more than a generation has passed away since that expectation was uttered, and still allotments of a proper size are far from being general. Unfortunately there is no means of accurately gauging the number of acres that has been so allotted. The Royal Commissioners on Agricultural Employment in 1868 estimate, however, that out of 320,855 acres, enclosed since 1845, and over which, common rights being unstinted, the general Inclosure Act of that year required that fair compensation should be made to the labouring poor by means of public allotments, only 2,119 acres were so assigned. Commenting upon this fact, the Commissioners state that this very inadequate precaution to secure the rights of the smaller commoners affords the greater reason why the owners of those inclosed lands should take care that the labouring poor should be in possession of field allotments of suitable size ; and further suggest that an annual return should be made by the officers of the Inland Revenue Department, by which it would be possible to ascertain how far the allotment system is carried out, and the average under that mode of cultivation " corresponds with, or falls short of the quantity necessary to afford that desirable accommodation and resource, in ample measure, to the labourers in agriculture."

Four years, however, after the publication of this Report by the Royal Commissioners, that great movement, under the leadership of Joseph Arch, began, which has done so much for the future of the English agricultural labourer. Not the least of the benefits which it has won for him has been the Allotment Extension Act, passed last year. The history of that measure is both interesting and instructive. I am indebted for the following account of it to my friend Mr. Howard Evans, who was in fact the real author of the Act. In the early years of the agitation, Mr. Evans had himself seen in the various parts of the country which he visited as a delegate of the Union, how greatly such an Act was needed. Here, however, are his own words. I quote with his permission from an unpublished paper :—

" Let me pay honour where honour is due. This question was first stirred by Mr. Theodore Dodd, the son of an Oxfordshire clergyman. Under the signature of 'Equitas' Mr. Dodd wrote in the Labourers' organ a series of articles on the local rights of farm labourers, which were afterwards widely circulated in a separate pamphlet. To the honour of some old-fashioned clergyman, Mr. Dodd pointed out that the men might elect a labourer as churchwarden if they chose, and in several cases this was actually done, though in one instance we had to serve a mandamus upon the Archdeacon before he would admit the labourer to office. Mr. Dodd called attention to the Act of William IV., which provided that fresh allotments should be let out to the labourers as gardens, and to the duty of the local authorities to put this Act in force. He held that the words of the Act applied not merely to fresh allotments under enclosure awards, but to all local charity lands. As many of the men had no allotments at all, and many others were allowed as a great favour to rent very small plots at fancy rents of from £4 to £8 an acre, there was much inquiry about local charity lands. In almost every county the District Secretaries of the Union obtained the county charity reports, and the clamour for the possession of the charity lands became general. In many cases applications were made to the trustees of the lands, who invariably treated these applications with indifference. Of course the next step was to apply to the Charity Commissioners. We were not then so fully aware as we are now of the fact that the Charity Commissioners delight in drawing up schemes which provide for the confiscation of the ancient charities of the poor in order that the property may be devoted to the benefit of the middle class. We did not want new schemes, we only desired that the Charity Commissioners should construe the old Act of William IV. as liberally as possible, and should use their influence with local charity trustees to secure its more complete enforcement. It was a very modest request, seeing that it would have injured nobody, and would have been of considerable benefit to the labourers ; but the Charity Commissioners would not allow me to appear as the representative of the men, and utterly refused to put upon the Act a liberal construction. In one or two cases, where we at length obtained an inquiry into the whole of the local charities and a new scheme was drawn up, we succeeded in making the transfer of the land to the men a part of the scheme. I recollect one case where there were only a few very small allotments at high rents a mile away, while there were 20 acres of charity land close to the men's own doors, which were let to the largest farmer in the parish. A few months after the labourers came into possession, I visited the village, and saw the men at work on their allotments in the evening. Some of them told me that their plots were as good as 3*s.* a week extra wages to them.

"When the Charity Commissioners showed me the door, the only course was to bring in a Bill which should make it clear that our construction of the old Act was to be acted on. I talked the matter over with Mr. Dodd, and then consulted a friend of mine, a barrister in the Civil Service. I explained to him that I wanted a Bill which would cover all local charities,—an exception was afterwards made of those left for church and educational purposes,—and he drew a Bill on the lines indicated. I then went to Sir Charles Dilke with the draft of the Bill, and he promised to introduce it, if I would collect the information necessary for making a speech in the House. I visited several counties for that purpose. And here let me pay a tribute to that fine old Tory gentleman, Mr. Henley, late member for this county. I found on the borders of Bucks and Oxon a number of allotments of good size, which had been let by him to labourers for several years. The arrangement was satisfactory to both parties, for although Mr. Henley let the land at the low rent of 25*s.* an acre, it had been previously let at not much more than half that amount. I saw the collector's book, from which it appeared that the rents were invariably paid. In other parishes I found miserably small allotments let at rents of £4, £6, and £8 an acre; in some villages I found the people packed together as close as in a London court, with no gardens at all. Sir Charles Dilke was supplied with a number of cases where the Act was badly needed, and in due time he moved the second reading of the Bill. The Conservative Solicitor-General ridiculed the measure, and Sir Charles Dilke only obtained ninety votes. He re-introduced it a second time, with no better results, and as farther progress was hopeless during the existence of the late Parliament, the Bill was dropped. After the General Election of 1880, Sir Charles Dilke became a member of the Ministry, and Mr. Jesse Collings, one of the oldest friends of the Union, took the matter up. The Bill passed with a general chorus of approval from both sides of the House of Commons, and the ex-Solicitor-General no longer denounced it as ridiculous. The Bill then went up to the Lords, who of course did their best to spoil it. It may be that some ardent Conservative here may be inclined to say that I am just like Mr. Bright, who, as Lord Norton says, cannot even make a speech on Temperance without girding at the Lords. But I have good reason for the complaint. I was so busy at the time that I did not watch what was going on; but as soon as the Act was printed, I looked it over and could hardly recognize my own child. It was as if a Chinese Lord and a Red Indian Lord had acted as its foster-parents; the one had cramped its feet with tight bandages, the other had flattened its nose with a board. The sweet simplicity of my Bill had disappeared, and the Lords had done their best to make its operation as difficult as possible. I had proposed to give the labourer a cheap and easy remedy in the nearest county court by summoning the trustees to show cause why the Act should not be put in force. At the expense of the loss of a day's work, and an outlay of 2s. or 3s., the matter would have been settled. The Lords struck out everything that referred to county court jurisdiction, and referred the aggrieved labourer to the Tite Barnacles of that abominable Circumlocution Office at Gwdyr House. The best of the joke was, that the Charity Commissioners had actually opposed the Bill; and the very Bill which they opposed they were called upon to administer. If these gentlemen had done their duty to the poor seven years ago, they would have made the Allotments Extension Act unnecessary. I am afraid that they are not likely to do their duty even now. Mr. Jesse Collings wrote me only a few days ago, 'The real difficulty is with the Charity Commissioners, who are not friendly to this Act; they will have to be fought seriously ere long.' The quantity of land coming under the provisions of the Act amounts to nearly a quarter of a million acres, but even now, owing to the action of the Lords, we are left very much at the mercy of the local trustees.

"The tricks resorted to by some of the trustees are simply infamous. In some cases they have let the land on a long lease so as to evade the Act, in others they have, contrary to law, charged exorbitant rents; in others they have, contrary to law, refused to let except to farm labourers, and sometimes only to farm labourers who are householders; in others they have ignored the Act altogether; in others they have illegally demanded half a year's rent in advance. In some of these cases the men have appealed to the Charity Commissioners in vain. So many complaints have flowed in to Mr. Jesse Collings that he has been compelled to issue an appeal for assistance in enabling him to start a temporary society called the Allotments Extension Association, which he calculates will be able to force the

4

hands of the Charity Commissioners and advise the men in
the various parishes, so that in two or three years the Act
shall be generally enforced."

In conclusion it may perhaps be useful to
give the main provisions of the Act :—

Section A.   All Trustees . . . of lands vested . . . for
the benefit of the poor of any parish

" (1) Shall set apart for the purpose of this Act such field
or other portion of the said lands as is most suitable, as
regards distance or otherwise, for allotments, and give public
notice, in manner directed by the Schedule of this Act, of
the field or portion so set apart, specifying the situation and
extent thereof, and the rent per acre or rod which they are
ready to accept for the same when let in allotments, and the
times and places at which applications for allotments are to
be made.

" (3) If the whole of the field or the portion so set apart
is let in allotments, the Trustees shall proceed, as soon as they
have power so to do, to set apart another field or portion of
their lands for the purpose of this Act, and give public notice
thereof as directed by this section, and so on until the whole
of their lands are let in allotments, or no applications are
received for further allotments.

" (5) If any of the said lands shall be found to lie at an
inconvenient distance from the residences of any cottagers or
labourers, it shall be lawful for the Trustees to let such lands,
or any part thereof, for the best rent that can be procured
for the same, and to hire in lieu thereof for the purposes of
this Act other land more favourably situated for allotments
to the poor of the parish or place for whose benefit such
lands are held in trust."

NOTE TO CHAPTER II.

Allotments Act of 1882.

Although this Act was passed for the benefit of the Labourers of the Country Districts eight years ago, very little benefit as yet has resulted from it. In many instances, unfortunately, the trustees of charity lands, even when they fully know their duties under the Act, refuse to carry them out, or to take any steps to let the Labourers and others in their districts know anything about the benefits to which they have a right.

I should strongly advise the inhabitants of any place where this difficulty exists to apply to the Secretary of the Allotments and Small Holdings Association, 12 Cherry Street, Birmingham, or to the Secretary of the Rural Labourers' League, 95 Colmore Row, Birmingham, both of which Societies have been established with the distinct object of helping labouring men to secure their rights under the Allotments Acts. If, however, they would prefer, in the first instance, to act independently :—

The following is the Form of Application to Trustees for Allotments, as prescribed by Act of Parliament :

ALLOTMENTS EXTENSION ACT, 1882.

To the Trustees of __ _____ Charity.

Parish of _____

County of_____

We, the undersigned, being Labourers or Cottagers resident in the Parish of_____in the County of_____entitled under the above Act to rent Allotments of the Lands of the said Charity, and being desirous of renting such Allotments, do respectfully require you, the Trustees of the said Charity, to give Public Notice of your intention to set apart the Lands of such Charity, or a portion of them, and to proceed to set aside such Land, as is required by Section 4, of the above Act.

Signed,

Allotments Act, 1887.

This Act of Parliament was passed in 1887 with the expressed purpose of making the local Sanitary Authority provide the labouring classes with allotments at a fair rent. The following summary of a Government return of the number of Allotments and Small Holdings, will show that some progress is being made, although it may well be doubted whether the increase in the number of Small Holdings is traceable to the provisions of the Act. The direct action of public opinion upon individual landlords seems to me, I confess, much more likely to have been responsible for the change. Certainly in no case have powers for the compulsory purchase of land been conferred on the Sanitary Authorities by the Local Government Board for the purposes of the Act. There have been two instances in which petitions for such power were declined. There have been five cases only in which loans have been sanctioned for the purchase of land. These, surely, are but miserable results to show for all the time and labour spent on this Act, and must sadly discount any satisfaction which the following figures from the parliamentary paper might seem to justify.

It appears that there are 455,005 separate detached Allotments now existing, and 409,422 Small Holdings other than Allotments. These figures show a very large increase in the number of recorded allotments, and an increase also, although less remarkable in extent, in the number of small holdings. The growth of allotments may be estimated by the following figures referring to the three periods named: 1873, 246,398; 1886, 357,795; 1890, 455,005. It will be observed that the rate of annual increase in the last four years has been apparently three times as rapid as between

1873 and 1886. In the report which precedes the return, Mr. P. G. Craigie, Director of the Statistical, Intelligence, and Educational Department, points out that considerable local variations occur in the proportions of allotments of a quarter of an acre or less, and allotments between that size and one acre. On the whole, the smaller class outnumber the larger by much more than two to one. But in particular counties the proportions are reversed. While in England, as a whole, there are 310,698 allotments under a quarter of an acre, and only 130,326 above that limit, in Bedfordshire, Huntingdon, Norfolk, Suffolk, and Worcester the larger type of allotments prevails, and in the East Riding of York, Lincolnshire, and Cambridge the allotments exceeding a quarter of an acre are nearly twice as numerous as those below that area. The counties showing the largest number of allotments are Northampton with 26,229, Wiltshire with 23,723, Leicestershire with 23,396, and Nottingham with 21,253. An examination of the details now supplied indicates that many allotments are in urban parishes, and presumably occupied by artisans. The mining counties of Durham and Glamorgan show a remarkable increase since 1886, their allotments appearing to have been more than doubled in the last four years. Large increases also appear in Kent and Stafford. In Devon, Essex, and Warwick comparatively little change is reported. In only four English counties is any decline apparent. In Cornwall the collectors ascribe the diminution as partly due to the removal of country labourers to more remunerative railway work. In Hereford, Northumberland, and the East Riding of Yorkshire there appear to be also fewer allotments, some of those returned in 1886 having been improperly included under that title, while in other instances plots then occupied as allotments have been required for building and other purposes. Adding to the 409,000 small holdings and 455,000 separate allotments, the special allotments granted by railway companies, so-called garden allotments attached to cottages, and other cases, it is estimated that in one form or another *petite culture* is in existence in Great Britain in at least 1,300,000 separate instances.

Summary of the Act, 1887.

Notwithstanding the unsatisfactory character of the Act in many particulars, it may be convenient that I should give here a brief description of its chief provisions.

(Section II.) Any six registered parliamentary electors or ratepayers living in the district, in the case of towns, or living in any parish of a Union in the case of the country, may bring the question of the need for allotments before the Sanitary Authority, which in rural places is the Board of Guardians. If the Sanitary Authority are of opinion that there is a demand for allotments, and that they cannot be got on reasonable terms by voluntary arrangements, they are directed, by purchase or hire, to get suitable land, and to let it in allotments to persons belonging to the labouring population. This section gives the opportunity for holding public meetings on the subject, and in other ways calling notice to the need for allotments.

It will not fail to be seen that this Section II. of the Act makes it impossible to compel the Sanitary Authority to do anything to provide allotments unless they are themselves in favour of doing so. The working classes, who were to be benefited by the Act, have no power over the Guardians, and here, on the very threshold of the Act, is plain proof of inefficiency for compelling the authority to carry out the wishes of those who stand in greatest need of land on which to grow food for their families.

The Act does not apply to shopkeepers, assistants or clerks.

Any land got for allotments must pay rent enough to cover all expenses which have been incurred in procuring it, of course including lawyer's bills, fencing, and compensation to landowner, &c., of which we learn more in the next section.

(Sections III., IV., and V.) These deal with the buying of land by the Sanitary Authority, and making it ready to let for allotments. When a Government is in a hurry to get a measure passed, another and older Act of Parliament is very often lugged head and shoulders into the new Bill. In the Allotments Act, therefore, it is ordered that in buying land, if any authority decides to do it compulsorily, the Land Clauses Consolidation Act, which is a long Act, passed years ago for the benefit of landowners, shall form part of the Allotments Bill; and the reason is pretty plain when we learn that under this Act a landowner can claim 10 per cent. over the value of his land taken compulsorily for disturbance, and beyond that it is possible for him to get 15 per cent. more for severance, or separation of one part of his land from another. With this clause staring them in the face there will evidently be few authorities wishful to buy land for allotments; and certainly it seems something of a pretence to speak of the Act as passed with the view of making it easier for the working classes to get allotments, when at the same time land may cost, to begin with, one quarter more than it is worth? After all this, the Local Government Board must send some one down from London to be sure that the land is wanted, etc.; and before it can be got for allotments an Act of Parliament for this special plot of land must be passed through both Houses of Parliament, and anybody may oppose the proposal who likes to do so. After the land is got, these clauses give power for it to be laid out.

(Sections VI. and VII.) One quarter's rent *may* be demanded in advance. The Sanitary Authority *may* make and alter rules for letting, as to whom shall be tenants, the size of the allotments (which in no case can be over one acre), the rent to be paid, etc. These clauses also forbid any building, other than a tool-house, shed, greenhouse, fowl-house, or pig-sty, to be built on any part of an allotment. The rules made must be confirmed by the Local Government Board in London before they can be issued. This is a striking instance of how little faith Parliament could place in the good intentions towards the working classes of the Sanitary Authorities. A further innovation on the land laws is made also by this clause, in the leave given to the tenant of an allotment at the end of his tenancy to remove any trees or bushes which he has planted, and for which he has no claim for compensation under the Compensation Act, unless he has obtained his landlord's consent to plant them.

(Section VIII.) We have here the question of arrears dealt with; and power is given to the Sanitary Authority to recover possession from a tenant who does not pay his rent within forty days, and if he does not keep to the rules made by the authority, or if he lives more than one mile out of the parish the tenancy may be ended at one month's notice.

The Bill gives an out-going tenant under this section a right to compensation, an arbitrator being appointed who will see that the Sanitary Authority deals fairly with an allotment tenant ; and, further, the allotment may be withheld from any fresh tenant until the compensation due to the out-going tenant is paid.

(Section IX.) In any place where allotments have been provided under this Act one-sixth of the electors may require the Sanitary Authority to hold an election to elect allotment managers. The election would be by ballot, and is directed to be held in any schoolroom which is partly supported out of the taxes, and the poorest man in the parish has as much right to vote, and his vote will be worth as much as that of the largest ratepayer.

The section, as will be seen, admits, so far as the machinery of elections is concerned, the thin end of the wedge of a good many things for which working men are fighting, such as *one man, one vote, the ballot,* and the right to use the parish schoolroom.

(Sections X. and XI.) In these sections directions are given as to the expenses incurred by the Sanitary Authority in dealing with this Act. The procedure under the Act is so complicated and expensive as to render it almost certain that when the expenses come to be charged on the rents of the allotments, they will be so dear as to be worth nothing. Section XI. gives power to the Sanitary Authority to dispose of surplus or unsuitable land.

(Section XII.) The Act lays down that an allotment may not be more than one acre of arable land, but by this clause the Sanitary Authority may establish a common cow pasture, on which those who have a cow could turn her out to graze by payment of a rent as for an arable allotment. It is an unhappy scheme. To have, say, an acre allotment in one place, and a cow pasture in another place, and the working-man's home in another place (*besides which no cow-house may be put on an allotment*), will daily, by its inconvenience and worry, remind a man of the blessings of legislating in a hurry.

Sections XIII. and XIV. permit Charity Trustees to hand over their land to the Sanitary Authority, and allow two or more parishes to unite together for the purposes of the Act.

(Section XV.) It directs the Sanitary Authority to keep a register showing the particulars of the tenancy, acreage, and rent of every allotment, whether let or unlet; and every ratepayer may, without paying any fee, examine this register and take copies of any part of it; and within a month of 25th March in every year the Sanitary Authority is directed to publish an account showing the outlay and receipts under the Act.

Sections XVI., XVII., and XVIII. conclude the Act, and are made up for the most part of explanations of earlier parts of the Act.

The following newspaper cutting will be useful as the record of what is likely to be a typical case.

Award under the Allotments and Cottage Gardens Compensation for Crops Act.

Mr. James Walker, one of the Leeds Borough Justices, has sat at the Town Hall, Leeds, as arbitrator to settle a claim under the Allotments and Cottage Gardens Compensation for Crops Act, 1887, made by George Eccles, of Farnley, against Benjamin Downs, of Farnley, farmer. Mr. Peckover, solicitor, appeared in support of the claim. The garden is rather less than one rood in extent, and the tenant, after occupying it for nearly two years, gave up the possession pursuant to a notice from his landlord on the 30th June. The claim amounted to £8 16s. 6d., including 10s. for manure and 30s. for labour, but those two items were disallowed by the arbitrator, who regarded them as inadmissible when a full claim was made for the resulting crop. Conflicting evidence was given as to the value of the crops left on the 30th June, when a portion only had been gathered, and the delay in this case showed the desirability of a prompt reference of such questions to some competent person. The landlord admitted that though served with a notice to appoint an arbitrator on the 10th July, he had neglected to take any steps. He now made a counter-claim of 40s. for weeds, and 14s. for dilapidations to a greenhouse. He also said he had re-let the garden without receiving any consideration for the crop, which he estimated at rather less than £1 in value. The neglect of the landlord to concur in the appointment of an arbitrator led to an application to the justices, who appointed Mr. Walker, and he of course acted without any remuneration. The award has now been published, giving £4 as compensation, and £1 6s. 10d. for costs, including solicitor's fee. This is the first application that has been made to the justices under the Act. As showing the productiveness of such gardens, it may be mentioned that one of the witnesses for the defence deposed that in a good year this small garden would yield as much in value as £20.

CHAPTER III.

" Parson do preach and tell me to pray,
    And to think of our work, and not ask more pay ;
    And to follow plough-share, and never think
    Of crazy cottage and ditch stuffs' stink—
    That Doctor do say breeds ager and chills,
    Or worse than that, the fever that kills —
    And a'bids me pay my way like a man,
    Whether I can't or whether I can ;
    And as I h'ant beef, to be thankful for bread,
    And bless the Lord it ain't turmuts instead ;
    And never envy the farmer's pig,
    For all a'lies warm, and is fed so big,
    While the missus and little 'uns grows that thin,
    You may count their bones underneath thar skin ;
    I'm to call all I gits the ' chastening rod,'
    And look up to my betters and then thank God."
                                          Punch.

IN the concluding sentence of the first chapter
I ventured to speak of the Social Destination
of Property. This is a doctrine which it
has often seemed to me to be specially
incumbent upon the clergy of Christ to
practise as well as preach. We country clergy,
at any rate, ought not to forget that we are, in

the majority of cases, if not landlords, at least landholders. Our numbers are estimated, I believe, at something like 1,200 in England alone. We represent, in fact, probably considerably more than one-fourth of the resident landowners of the country. We have then, it seems to me, a special duty with regard to this relation. Can we do nothing then, I would ask, in our character of Glebe Landlords, to mitigate one at least of the great evils arising from absorption of small holdings? Would it not be possible for us to use the property of which we are trustees in such a way that we should once more be able to set before the rural labourer "that one attainable point of hope," without which I venture to say all plans for his welfare, whether social, economic, or religious, will be in vain? The root virtues of a manly character—self-reliance, self-help, independence, ambition—will never grow in the hard soil of a day labourer's life, unless they be first watered from the perennial spring of hope. To transpose the old proverb, "While there is hope there is life." Let us find a means of implanting this principle in the bosom of the agricultural labourer, and we may succeed

in changing even the dull grey monotony of his
existence into one of—

"Life, full life,
Full flowered, full fruited, reared from homely earth,
Rooted in duty, and through long calm years
Bearing its load of healthful energies,
Stretching its arms on all sides, fed with dews
Of cheerful sacrifice and clouds of care."

Let me indicate briefly, then, three practical
directions in which I think it is open for the
clergy to do something in this matter :—

Apportionment of Glebe Allotments.

In the first place, I should consider that
where allotments of sufficient size do not already
exist, it is the duty of the holder of the Glebe
to divide a portion of that land, with the
object of accepting his labouring parishioners
as tenants. If I may venture to quote my own
experience, I am sure he will find few acts
of his parochial administration upon which he
will look back with more sincere and unmixed
satisfaction. It will be necessary, of course,
to protect himself by proper rules and precautions.

But, after all, no rules or precautions are so
important as the precaution of securing in the

first place a good state of feeling between parson and men.

Village Parliaments.

I may perhaps be allowed to mention here one means of securing this, which I have myself found of the very greatest use? When I first thought, some years ago, of dividing my Glebe land among the labourers, I happened to hear of certain experiments in co-operative farming which were being carried out at Blennerhasset, in Cumberland, by a brother of Sir Wilfrid Lawson. One feature in his scheme struck me as extremely original. He was in the habit of periodically assembling all the workers on his farms, for general purposes of consultation, in a sort of open council, which afterwards came to be known as the Village Parliament. The idea of inviting miscellaneous criticism in this way was certainly droll, but it is not perhaps, after all, so foolish as it looks. At any rate, as a result, I determined to invite my labouring parishioners to meet at my house once a week, to talk over questions of agricultural economy. For several years we were accustomed to meet, during the winter months, at my house in this way. In the first instance, the proposed division of the Glebe land suggested to us plenty of topics for conversation. Subsequently subjects for discussion were found in such books as Mr. Brassey's "Work and Wages," Thornton on "Labour" and " Peasant Proprietorship," Kinnaird Edwards' " Rural Economy," Holyoake's " History of Co-operation," and Mr William Lawson's "Ten Years or Gentleman Farming," passages from which I was accustomed to read to them.

I do not know whether the labourers have learnt very much from me, but I can honestly say that I have learnt very much from them. Again, there is another possible way in which perhaps, the country clergy might help the rural labourers.

Improved System of Small Husbandry.

Mr. Gladstone, in a very admirable speech on garden cultivation, which he delivered some years ago at Hawarden, drew attention to one important method by which he considered that the cultivators of the soil in this country might very materially improve their position. He repeated that advice in a speech to the electors of

Midlothian during his celebrated campaign in 1880. After describing the peasant properties of France, he asks, "What do these peasant properties mean? They mean the small cultivation, that is to say, the cultivation of superior articles on a small scale—cultivation of flowers, cultivation of trees, cultivation of shrubs, and cultivation of fruits of every kind—all that, in fact, which rises above the ordinary character of farming production, and rather approaches that of gardening." And he goes on to express a belief that a "great deal more attention will have to be given than heretofore to the production of fruits, vegetables, and flowers, of all that variety of objects which are sure to find a market in a rich and wealthy country like this, but which have been confined almost exclusively to garden production."*

---

* See also on this point a most excellent pamphlet, full of sound practical advice, on "Market Gardening for Farmers," by Mr. Charles Whitehead, of Maidstone, reprinted from the *Mark Lane Express:* "If the conditions of soil and climate are generally favourable for the successful production of vegetables, what has prevented farmers, and what does prevent farmers, from going into the business? Chiefly, it must be answered, the conservatism which clings so tightly to them, and binds them to old customs and the traditions

Now, there is no doubt, I should imagine that the system of husbandry generally in vogue on labourers' allotments might be very advantageously improved in this direction. Probably no use to which the soil can be put is more profitable than that of market gardening. Unfortunately any attempt at improvement in this direction, except, of course, in the vicinity of large towns, is usually considered to be useless, owing to the want of convenient markets. Now, ought this to be an insuperable difficulty? In these days surely, when the railway system penetrates into almost every corner of the

---

of mediæval methods ; a conservatism, by the way, which has been somewhat modified of late by the action of foreign competition, and by the falling away of the scales from many eyes." Contrast with this "conservatism " of English farmers the "go-ahead " character of these same foreign competitors, as reported by Mr. Sewell Read and Mr. Pell to the *Agricultural Interests Commission* two years ago : "It is pleasing to notice the willingness, it may be even called ready eagerness, with which the American farmers welcome all things new. Any novelty, however revolutionary to existing plans and ideas, is sure to find admirers, and may expect a fair trial. The rapidity with which new systems of daily management have spread would astonish many English farmers, who, especially in dairy districts, are proverbially slow to change."

5

kingdom, the question of markets is, after all, merely a question of organization.*

Co-operative Distribution of Garden Produce.

I would venture to suggest, therefore, that the question of the distribution of the produce of garden allotments offers a most admirable opening for the application of the co-operative principle. Would it not be possible in many cases, I would ask, to associate the allotment tenants in some form of co-operative society, which should have for its object, not perhaps, in the first instance, the working of the allotments in common, but for making such arrangements with the Railway Companies as should enable them to get within reach of the best markets for the distribution of their produce ?

We are only at the very beginning, as it appears to me, of the application of the principle of co-operation to agriculture. If the country clergy will but endeavour to help their labouring parishioners to apply the principles of

---

* A very practical comment on this statement will be found in the account of the new market opened by the Great Eastern Railway, given at page 155, Appendix. Cf. also Mr. Whitehead's pamphlet mentioned above, p. 52.

co-operation to the conditions of rural life, I venture to think that not only will there be few acts of their parochial administration upon which they will be able to look back with more sincere and unmixed satisfaction, but there will be few also in which, in the light of another world, they will seem to themselves to have been more directly " feeding the flock of Christ committed to their care." The story of self-denying enthusiasm and noble endeavour, which I shall endeavour to tell in the following chapters, will, I hope, do something to show how co-operation rightly understood is but the endeavour to realise in economic life the social ideal of Christianity.

CHAPTER IV.

CO-OPERATIVE SMALL FARMING.

" Without attempting to predict the exact phases through which co-operation will pass, it can scarcely be doubted that the principle is so well adapted to agriculture, that it is certain some day to be applied to that particular branch of industry with the most beneficial results. . . . The progress towards co-operative agriculture will no doubt be slow and gradual. The labourers will have to advance towards it by many preliminary steps."—RT. HON. HENRY FAWCETT, M.P.

IT is always a matter of some surprise to those who make a study of the history of Co-operation, that so little progress should have been made in the application of the co-operative principle to agriculture,—a surprise which is only increased when it is remembered, on the one hand, that the earliest form of landed tenure was that of agricultural association through community of land, and, on the other, that in few undertakings would the conditions of co-operative success seem to be more conspicuously present than in that of farming. When, therefore, in reading of the vast development of co-operative enterprise in England during the last half-

century, we find that the number of co-
operators has increased from 28 in the year
1844 to 526,000 in the year 1880, and that the
capital invested during that period has increased
from £28 to very nearly £6,000,000, it does
seem very discouraging to those who have been
inclined to regard co-operation as a possible
regenerating force in modern economic society,
to find that, in connection with the agricultural
classes at any rate, who above all are in such
need of social and economic improvement, co-
operative enterprise should seem to have made
hardly any headway at all during the last fifty
years.

From time to time, co-operators themselves
become alive to the fact as conveying somewhat of a reproach to their energy and faith.
At both the last annual Co-operative Con-
gresses, prominent expression was given to
this feeling. At Oxford last year, a very able
and suggestive paper was read on this subject
by Mr. Kitchin of Christ Church, now Dean
of Winchester, in which, after showing how,
in his opinion, co-operation might best grapple
with the chief problems of farming, after indi-
cating the unusual facilities for the work, and

the specially bright hopes of success in it, he concluded by commending the subject with its far-reaching consequences to the mature judgment of those who had already proved their courage and administrative ability by bringing co-operative organizations to so successful an issue.

This paper was fully discussed at the time, and was afterwards, by resolution of the Congress, referred to the consideration of the district conferences. The reports of these discussions have from time to time appeared in the pages of the *Co-operative News,* and have no doubt served to create considerable inquiry on the part of working co-operators as to the possibilities of co-operative agriculture.

In the concluding remarks of his inaugural address at this year's Co-operative Congress at Edinburgh, the President, the Right Hon. W. E. Baxter, M.P., again insisted upon the importance of co-operators giving their attention to the subject of co-operative agriculture.

It is my object, however, in the present pages to press the importance of this subject, not only upon the general attention of co-operators, but also, as far as I may be able, upon the special attention of agricultural labourers, who would themselves evidently be the class most especially benefited by the success of co-operative enterprise in this direction.

I am well aware of course that the back-ward condition of the rural labourer, as compared with other sections of the labour class, has often been advanced as one of the chief difficulties in the way of the successful application of the associative principle to agriculture, and it is true, no doubt, that his narrow school and social education does unfit the agricultural labourer to some extent for co-operative work. At the same time this is hardly a reason, as it seems to me, why co-operators, at any rate, who are generally somewhat proud of claiming that their movement is one quite as much for the promotion of the general well-being as for the protection of individual interests, should hesitate to encourage co-operative enterprise among agricultural labourers. If it is indeed the want of a wider social organization which has mainly stood in the way of the extension of the principles of co-operation to agriculture, a true appreciation of the value of the co-operative faith would certainly seem to suggest that some organised effort should be made to meet that want. A co-operative faith implies a co-operative propaganda.

Indeed, for my own part, I cannot but think that such propaganda is a first necessity before any reality can belong to the discussion of large schemes for the establishment of co-operative farms, such as that, for example, suggested in the otherwise admirable paper of the Dean of Winchester.

That the energy and enterprise which has made the success of some of the great distributive societies in the north of England is equal also to the organization of a successful co-operative farm, I have no difficulty in believing. As Dean Kitchin truly says, " Co-operators have shown by the business capacity of their organizations that the directive power is already there." I cannot, however, quite so readily as he appears to do, bring myself to believe that " the adaptation of this ability to the management of land is a *mere* matter of detail." It is a matter of detail, no doubt, but of detail which, if the farm is to be anything further than a Joint-stock association, must depend very largely upon the spirit in which the working labourers of the farm are able to appreciate and intelligently to carry out the principle of co-operation.

For this reason, therefore, I cannot but think that to propose in the first instance the organization of large co-operative farms, worked with the capital and "engineered" (to use a convenient American phrase) by the business managers of some of the larger existing co-operative societies, is in reality to begin at the wrong end, and, moreover, entirely ignores the lesson which the history of co-operation ought to teach. The enormous success of distributive co-operation has been built up, it should never be forgotten, from very small beginnings, and has been a matter of slow and gradual growth. So also in all probability must it be with agriculture, if co-operative farming is to be equally successful. The twenty-eight "poor labourers" of Rochdale will have no doubt their agricultural after-types. For this reason, therefore, it seems to me every way better that those who are desirous of seeing the development of co-operative agriculture, should confine themselves for the present, at any rate, rather to the propagation of co-operative truths among the rural classes, and the encouragement of such efforts on the part of the labourers themselves, however humble they may be, which would seem to be the natural and spontaneous result of such propaganda.

In the following chapters I have endeavoured to give as plain and simple an account as I could of some of the more interesting and remarkable experiments in co-operative farming, which seem to me at all likely to furnish useful material, either by way of warning or of encouragement, for such a propaganda as that I have indicated.

CHAPTER V.

"Siouan-wang, the King of Tshi, said to Meng-tseu, 'I have been told that the park of the King Weng-wang was seven leagues in circumference; was that the case?' Meng-tseu answered respectfully, 'History tells us so.' The King said, 'If so, was not its extent excessive?' Meng-tseu answered, 'The people considered it too small.' The King said, 'My insignificance has a park only four leagues in circumference, and the people consider it too large; whence this difference?' Meng-tseu answered, 'The park of king Weng-wang contained all these leagues; but as the King had his park in common with the people, the people thought it small. Was that wonderful? I, your servant, when I was about to cross the frontier, took care to inform myself of what was especially forbidden in your kingdom, before I dared to venture further. Your servant learnt that there was within your line of customs a park four leagues round, and that the man who killed a stag there was punished with death, as if he had killed a man. So that there is an actual pit of death of four leagues in circumference opened in the heart of your kingdom. The people think that park too great. Is that wonderful?'"—CONFUCIUS.

OF all the social experiments that have yet been made in the direction of applying the principle of Associated Labour to the occupancy and tillage of the soil, the Co-operative Farm established fifty years ago by an Irish landlord, Mr. Vandeleur, at Ralahine, in county Clare, is by far the most interesting and instructive.

A Romance in Facts and Figures.

The story of this successful and suggestive experiment at one time attracted much attention from both economists and politicians in this and other countries, and, had it not been for the premature collapse of the undertaking, from a cause personal to the landlord, and in no sense affecting the principle or merits of the scheme, would no doubt have become a standing example of the great possibilities that lie open for Co-operative Small Farming in the future. The story has lately been retold by Mr. E. T. Craig, the organiser and first secretary of the association, an old man now, in a little book entitled, " The History of Ralahine and Co-operative Farming." It is published by Trübner and Co., price two shillings. I can strongly recommend it as a book of the greatest value not only to the co-operator, who will find it full of wise thought and noble sentiment on the subject of Associated Labour, but also to the perplexed politician who is seeking some solution to the many problems, social, agrarian, political, which are summed up for him in the one phrase, " The Irish Question." The book, moreover, is not only most instructive, it is deeply interesting. " If ever there was a romance in facts and figures,"—truly enough says the *Spectator*,—"it is the story of Ralahine, a fairy tale of political economy, by one who had been an eye-witness of its reality."

Ireland Fifty Years ago

Ireland we have all long known as a fit realm enough for romance, but our direful experience of the last few years may well have taught us to think that any romance there can possibly be to tell about Irish land must be one in which the bright pages would stand out against a very dark background indeed. And we shall not be wrong in such surmise. Fifty years ago, however, the background was even blacker than it is to-day. Look at this picture, which Mr. Craig draws of the state of Ireland and the Irish in the year 1830, when he first began to organize the Ralahine scheme :—

" The population of Ireland amounted to about seven

and a half millions. Poverty is marvellously prolific, even when want grows faster than food. Land in Ireland avail-able for tillage is limited, and as it forms the basis of existence, it becomes an object of great vital importance to obtain it, and hence competition had raised its value while it reduced the wages of labour and the means of subsistence. The food of the peasantry consisted chiefly of potatoes. In the south and west the crops had failed. To add to the evil, large landlords had begun to evict their tenants, and to reduce small holdings, while, owing to the want of capital and of confidence, tillage lands were converted into grazing farms, on which a herdsman and boy could supersede some twenty labourers and ploughmen. The rents of con-acre were both enormous and unfair, from the fact that the poor tenant supplied the manure for the potato crop, while the landlords took the benefit in the grain crop subsequently sown. Rents were demanded at £8, £10, and in some cases at £14 per acre. If the tenant could raise a sufficient crop of potatoes to pay the rent and sustain his family, he considered himself fortunate. In many cases the crops were taken to the market attended by the agents, and the proceeds handed to them in payment of rent, while the slave of toil returned home empty-handed, with the galling knowledge that the fruits of his labour were taken by another, who was perhaps the representative of an absentee. In bad seasons, famine soon became prevalent. The labourer and his family under such circumstances were doomed to want and starvation. Peace and order were impossible. Coercion Bills, Arms Bills, an armed Police Force of 30,000 men, and a large proportion of the British Army might make a solitude, but that would not make peace, order, and contentment. Under the conditions indicated, many perished in silence, while thousands, alike ignorant of the causes of their misery and of the remedy, banded together in the vain hope of finding a cure for their sorrows by striking terror into the great landlords, their agents, and the Government. They saw no way of life and existence for them, save through the meshes of crime and the bloody portals of force, violence, and murder."

Similar evidence to that of Mr. Craig is furnished by the "Annual Register" of 1831, which bears testimony to the severity of the social crisis through which Ireland was then passing :—

"The peasantry marched in bands through the counties, demanding reduction of rents and increase of wages, and threatening destruction to the magistrates and gentry who should disobey or endeavour to resist. . . . The serving of threatening notices, the levelling of walls, the driving off of cattle, the beating of herdsmen, the compulsory removal of tenants, the levying of contributions in money, the robbery of dwelling-houses, the reckless commission of murder, were driving the better classes of inhabitants to desert their houses and seek refuge in some other quarter."

Force no Remedy.

Such, then, was the dark and unpromising background upon which the first bright pages of the Ralahine Romance had to be written. While other landlords were flying in terror from this scene of outrage, murder, and lawlessness, far exceeding in extent and violence anything of recent occurrence, and leaving the armed police and the English soldiers to cope with men upon whose hearts

"Famine had written fiend,"

there was one Irish landlord at least brave enough to face the storm, and in faith that "force was no remedy" had the courage to set himself calmly to the task of seeing how far the principles of co-operation, which he had learnt from the great English Socialist, Robert Owen, would go towards a solution, on his own estate at any rate, of the Irish Land Question.

The Ralahine Farm.

The scene of the new social experiment was admirably adapted for the purpose. Ralahine consisted of 618 acres, about one-half of which was under tillage, with suitable farm buildings, and situated between the two main roads from Limerick to Ennis. A bog of sixty-three acres supplied fuel. A lake on the borders of the estate gave a constant and available supply of water power, and a small stream flowing from it gave eight-horse power to a thrashing mill, a skutch and saw mill, a lathe, and so forth. A fall of twenty-horse power was available at a short distance, when required for manufacturing purposes.

A large building was erected by Mr. Vandeleur, 30 feet by 15 feet, which should be suitable for a common dining-hall, with a room of the same size above, available for lectures, reading-room, or classes. Close to them he built a store-room, with a dormitory above. A few yards from and at right angles to the la͡ge rooms he put in course of erection six good cottages. Several hundred yards away stood the old castle of Ralahine, with lofty square tower and arched floors, capable of being temporarily adapted for the accommodation of those whom Mr. Vandeleur hoped to unite in his new system of mutual co-operation.

The actual site for the proposed Co-operative Farm was in fact all that could be wished—fair soil of sufficient extent, good water-power, abundance of cheap fuel, extensive buildings, hard roads, nearness to two market towns.

Irish ideas of Co-operation.

But what of the proposed co-operators?

6

That they could not have been very different from the rest of their fellows, of whom I have just written, the following incident will, I think, sufficiently show

Mr. Vandeleur's last steward had been somewhat despotic, harsh, and severe in his treatment of the labourers on the estate. A reaper on one hot harvest day had paused from his work to get a drink of water from his can, whereupon the steward kicked it over, declaring that he would not have water there as an excuse for the reapers wasting their time. Similar acts of harshness roused a spirit of revenge. A midnight meeting was held in Cratloe Wood. The steward was condemned to death. Lots were drawn as to who was to do the foul deed. A few nights afterwards, in the presence of his wife, to whom he had only been married three months, he was shot dead as he was bolting his door. The assassin escaped, and was never brought to justice.

Not very promising materials, one would think, out of which to form ideal co-operators. At least their present idea of associative labour was of a somewhat ghastly type !

At any rate it must be allowed that he must

have been a bold man who, with such a possible fate before him, was ready to undertake the
task of endeavouring to organize these *ci-devant*
" Whiteboys " and " Terry-Alts " into a civil-
ised community of co-operative farmers.

A Lancashire Lad,

Such a helper, however, Mr. Vandeleur was
fortunate enough to find in Mr. Craig, a man
not only, as we may well suppose, of rare pluck
and courage, but one capable also of bringing
practical skill, foresight, and perseverance to
the intelligent application of those principles of
co-operation in which he believed so enthusias-
tically. He was a native of Lancashire, one
of those " Lancashire lads," in fact, of whom
he exclaimed soon afterwards—when on his
first visit to Limerick he saw the neglected state
of the river Shannon, one of the noblest rivers
in the British Isles, with its splendid natural
resources for water carriage, so typically left
undeveloped by the Irish because, as they said,
they were waiting for " Government help "—
" Some Lancashire lads I know would have
made short and cursory work of waiting for
Government. ' Hang the Government ! Why

wait for them ? Let us co-op. and do the work ourselves !' " Lancashire was already begin- ning in those days to take that lead in the development of co-operative enterprise among the working classes which she has ever since so nobly retained (I see by the returns lately published in the Congress Report that there are in that county this year 159,478 members of registered co-operative societies, doing busi ness to an amount of over eight millions, with a net profit of £597,434), and which makes one hope that in this respect, even more perhaps than in some others, Lord Derby's words may be true—" What Lancashire thinks to-day England will think to-morrow."

A short tour through the part of Ireland where the experiment was to be tried, with the object of studying the character and condition of the people with whom he would shortly have to deal, soon brought Mr. Craig to Lord Bacon's opinion—" To allay sedition we must allay the makers of it."

" The conclusions to which the relations of the labourer to the land and the fruits of his toil led me were that the causes at work were social and agrarian, as well as political, and that

social amelioration, and a share in the net profits, if any, after paying rent and the interest
of capital, would realise a great change at once
in the spirit and ameliorate the condition of the
people."

Mr. Craig lost no time in preparing a draft
of the constitution and laws of the proposed
association founded on these principles. This,
after being approved by Mr. Vandeleur, was
submitted to the members and signed. No
alterations were found necessary during the
experiment. Want of space prevents me from
quoting in full the whole document. I give,
however, sufficient, I think, to judge of the
essential features of the scheme, and one or two
of the more characteristic rules :—

LAWS OF THE RALAHINE AGRICULTURAL AND MANU-
FACTURING CO-OPERATIVE ASSOCIATION.

Preamble.

The objects of this Association are :—

I. The acquisition of a common capital.

II. The mutual assurance of its members against the evils
of poverty, sickness, infirmity, and old age.

III. The attainment of a greater share of the comforts
of life than the working class now possess.

IV. The mental and moral improvement of its adult
members.

V. The education of their children.

Basis of the Society, &c.

1. For the attainment of the foregoing objects the persons who have signed these rules agree to associate together, and to rent the lands, building, manufactories, machinery, etc., of Ralahine, from Mr. John Scott Vandeleur, according to " agreement," and they each of them, jointly and severally, bind themselves to obey the following rules, and to use every means in their power to cause them to be observed.

2. That all the stock, implements of husbandry, and other property belong to and are the property of Mr. Vandeleur until the society accumulate sufficient to pay for them ; they then become the joint property of the society.

*   *   *   *

Production.

9. We engage that whatever talents we may individually possess, whether mental .or muscular, agricultural, manufacturing, or scientific, shall be directed to the benefit of all, as well by their immediate exercise in all necessary occupations as by communicating our knowledge to each other, and particularly to the young.

10. That as far as can be reduced to practice each individual shall assist in agricultural operations, particularly in harvest, it being fully understood that no individual is to act as steward, but all are to work.

11. That all the youths, male or female, do engage to learn some useful trade, together with agriculture and gardening, between the ages of 9 and 17.

12. That the committee meet every evening, to arrange the business for the following day.

*   *   *   *

15. That no member be expected to perform any service or work but such as is agreeable to his or her feelings, or

they are able to perform ; but if any member thinks that any other member is not usefully employing his or her time, it is his or her duty to report it to the committee, whose duty it will be to bring that member's conduct before a general meeting, who shall have power, if necessary, to expel that useless member.

*   *   *   *

Education and Formation of Character.

26. That each individual shall enjoy perfect liberty of conscience and freedom in the expression of opinions and in religious worship.

27. That we each observe the utmost kindness, forbearance, and charity for all who may differ from us in opinions.

29. That no gaming of any kind be practised by any member of the society.

31. That no spirituous liquors of any kind, tobacco, or snuff be kept in the store or on the premises.

*   *   *   *

Government, etc.

37. That the society be governed and its business transacted by a committee of nine members, to be chosen half-yearly by ballot, by all the adult male and female members.

41. That there be a general weekly meeting of the society ; that the treasurer's accounts be audited by the committee and read over to the society ; that "the Suggestion Book" be also read at this meeting.

A memorandum of agreement was also drawn up between Mr. Vandeleur and Mr. Craig and three other members of the Association, defining the conditions and terms of the letting of the farm, and making provision for increasing the rate of wages and dividing profits in the event of the experiment being brought to a successful issue. The arrangement was in brief this. The farm was let by Mr. Vandeleur at a fixed rent, to be paid in fixed quantities of farm produce, which, at the prices ruling in 1830-31, would bring £900, which included interest on buildings, machinery, and live stock provided by Mr. Vandeleur. The rent alone was £700. As the farm consisted of 618 acres, only 268 of which were under tillage, this rent was a very high one—a fact which was acknowledged by the landlord. All profits, after payment of rent and interest, belonged to the members, divisible at the end of the year, if desired.

It will be seen that the method of paying the rent (I quote again Mr. Craig's own words) differed from the old and accustomed methods. The prices ruling the Limerick markets in 1830-31 were taken as standard prices during the existence of the Association, for the six articles in which rent was paid, and it was felt · to be just, and gave satisfaction to both landlord and tenants. If the produce of the farm had increased, or say doubled temporarily by

the effect of an exceptional season, or permanently by improvements on the part of the
members, the Society would have appropriated
the difference. In the case of permanent
improvements, the landlord would of course
have been benefited by an increased value of
the property. In either case the increase would
have arisen from causes beyond the control and
quite independently of the landlord. It would
have arisen either from increased industry, care,
and skill, improved mode of tillage, increase of
the acreage under cultivation, or from an
unusually favourable action of nature's laws,
effecting a greater absorption of the various
elements of the earth and the atmosphere,
which go to form plant life. The proprietor
would not have supplied this extra industry,
these improved methods or extra forces in
nature's laws. . . . Under the arrangement
made with the proprieter, the society had the
full benefit of the skill, industry, and enterprise
of its members, and had the advantage of good
seasons; while, on the other hand, the landlord
reaped the advantage of any advance in market
prices which in course of time would result
from increased demand. · Had the Society

neglected the proper cultivation of the land, it would have risked having no surplus to divide among the members, and have been liable to risk the loss of its occupancy.

The Old and the New System.

Such then in effect are the main features of the Ralahine Scheme. The principles (as enunciated above by Mr. Craig) upon which " The New System," as it soon came to be generally called, was to be worked, will doubtless strike many of my readers as practically condemned in advance by the fact that they are entirely contrary to all the recorded canons of orthodox Political Economy. Well, I am very sorry—I cannot help it ; but I am tempted to say, " So much the worse for the orthodox Political Economy." Like many other forms of orthodoxy, the stress of nineteenth century human needs and life may have to teach even the " Science of Wealth " some new dogmas— among the chief of which I trust will be this, " In the last resort the question is not about wealth but about men."

What the relations of the landlord, the tenant farmer, and the labourer, both in Ireland and elsewhere, under the " old system " of competitive selfishness, have been, all the world knows. What those relations might become for landlord, tenant farmer, and labourer, both in Ireland and elsewhere, under the "new system" of mutual co-operation, the issue of the Ralahine scheme, may, I trust, demonstrate.

CHAPTER VI.

THE NEW SYSTEMITES.

" Of old things all are over old,
   Of good things none are good enough ;
   Let's try if we can help to mould
   A happier world of better stuff."

MR. CRAIG, " the Lancashire lad who had come
to teach Irish labourers how to co-op. and
do for themselves," by no means found everything plain sailing, we may well suppose, when
he set himself to put into operation the
scheme of co-operation which I described in the
last chapter. The labourers on the estate
were discontented, moody, and suspicious, and
at first Mr. Craig's appearance among them
only tended to increase their want of confidence. They regarded him merely as a " new
steward." He was a stranger, moreover, and
a " Sassenach.'

" Being the only Saxon in that part of
Ireland," he says, " and arriving while the
people were in a state of wild frenzy of

indignation against their forced exclusion from the soil, they naturally concluded that as traditionally all Sassenachs were incapable of dealing fairly and justly towards Irishmen, I should secretly sympathise with the landlords and the police authorities. Their prejudices and suspicions led them to suspect me as likely to betray them by obtaining the name of the man who had murdered the steward."

Irish Humour.

On one occasion he was cautioned not to return to his lodgings by the same road as that by which he left if detained after sunset. On another, he was struck with a stone ; and on yet another, he was presented with a sketch of a skull and cross-bones and a rudely-drawn coffin, with an intimation that they intended to put him to bed under the "daisy-quilt." Altogether he had a somewhat unpleasant time of it. But he never lost heart ; he had faith in the principles of the " new system " which he had come to establish, and he was not to be disheartened at the outset by difficulties and obstacles which he had partly foreseen.

He set himself accordingly to study the

character of the people in order that he might learn how to help them. He was not long in discovering that there was much that was very lovable about the Irish people.

" Kindly sympathy, tenderness, and hospitality," he says, " are marked characteristics of the native Irish. On entering the cabins of the district I found the manners of the people naturally easy at receiving a stranger. In England if a person enters a house or a cottage where he is unknown he is received with a stare of surprise and a scrutinizing gaze of doubt before he feels at ease. In the south of Ireland if the stranger goes into a cabin where he is unknown, he will be received with 'Caed mille failthe'—a 'hundred thousand welcomes,' or 'God save you kindly,' in return for the salutation of 'God save all here.' "

Mr. Craig's desire to return these kindly greetings in the same language in which they were uttered almost led him on one occasion into a somewhat unpleasant experience of the truth of the old adage that—

" A little learning is a dangerous thing."

He had requested one of the more intelligent

of the Ralahine labourers to teach him the true Irish form of salutation in answer to the kindly wayside greeting of "God be with you," and was instructed by his humorous, but as the sequel turned out 'rather malicious tutor to reply, "Tharah ma dhoel!" Here is an account of the incident—

"When a stranger addressed me with 'Dea vaha,' I replied 'Tharah ma dhoel!' If another said, 'Peace be with you,' I still replied, 'Tharah ma dhoel!' I observed that my civilities had a somewhat puzzling effect on the passers by the way. It occurred to me that I was too hurried or too indistinct. My next experiment, however, was a critical one. The wayfarer was a tall, sturdy son of the soil, with a long-tailed frieze coat, who carried a stout blackthorn stick or shillelah. He gave the usual civil recognition, and I promptly replied—'Tharah ma dhoel!' My reply seemed electrical The fellow stood stock still, and by a clever jerk he threw his shillelah up in the air and caught it in the middle, and then giving it a twirl, he said—'Say that agin, say that agin, and I'll lay ye in the turf-pit!'"

After this experience Mr. Craig thought it well to get a literal English translation of his Irish salutation, and finding that "Tharah ma dhoel!" in Irish means "Go to the devil!" he wisely changed his tutor for one less clever but more frank.

Progress of the Association.

During this time the various buildings before mentioned were rapidly approaching completion. On the 1st of November, 1831, everything was ready for a start. Accordingly the whole of the labourers and artizans on the estate and some living in the immediate neighbourhood were assembled, to the number of about forty, and the scheme of the proposed Association was fully explained. Mr. Craig, however, perceiving some uneasiness of feeling on the very threshold as to who were to compose the society, and believing that he himself was still regarded with suspicion by some, proposed that the election of the members and the officers of the Association should be by the ballot of the men themselves. This plan was adopted. From that moment the success of the scheme may be said to have been assured.

The numbers admitted at this first ballot were :—

Adult single men . , . . .	21	
Married men	7	
		— 28
Single women	5	
Married women	7	
		— 12
Total		40

Orphans under 17 years of age—

Boys	4
Girls	3
Infants	5
	— 12
Total (with adults) .	52

The total number afterwards increased to 81.

Ralahine awakens Hope.

In less than two months from this time the Association was in complete and satisfactory working order, and the comfort, freedom, and industry of the members soon began to attract general attention not only in county Clare, but in the whole of that part of Ireland. The "new system" was the talk of the people at their places of meeting far and wide. It was hoped that other landlords would imitate the excellent example of Mr. Vandeleur, especially as his example was one profitable to himself as well as calculated to produce peace and contentment in disturbed Ireland. It was this feeling and the evident wish for the continued success of Ralahine that caused the murders and agrarian outrages to cease in that part of the country. They did not occur again on the Ralahine property for more than thirty years afterwards,

7

The "old system," however, has again produced
the old fruits, and under the Coercion Act of
1881 county Clare once more became a pro-
claimed district.

The following testimony to the success of
the undertaking was given by Mr. Finch in
evidence before a Commission of the House
of Commons in 1834 :—

"I saw an agricultural institution in Ireland last year in
county Clare, all the arrangements and laws of which are so
excellent, and point out so clearly the certain means of removing immediately the ignorance, mendicity, pauperism,
drunkenness, and crime that exist in both countries, without
any extra outlay of capital or interference with existing
institutions, that I am determined to devote a considerable
portion of my time to the promulgation of them.   They are
most important to landholders."

A fair idea of the method and operations of
the Society will be gained from the following
statistics :—

ABSTRACT OF LABOUR SHEET FOR WEEK ENDING
JANUARY 14, 1832.

Farm.	£	s.	d.
Carting out and mixing compost manure	1	1	6
Ploughing in Granapan and Calf Field .	0	13	4
Washing and steaming potatoes . .	0	6	0
Threshing & preparing wheat for market.	0	18	0
Conveying wheat to Limerick . .	0	10	8
Pulling and drawing in turnips , .	0	8	0

ABSTRACT OF LABOUR SHEET (*continued*).

	£	s.	d.
Trenching in wheat	0	9	0
Attending and foddering cattle	0	16	0
Carpenter's labour on farm	0	7	4
Smith's ditto	0	4	8
Herding stock	0	5	0
Dairy	0	5	0
Manufacturing wool into frieze	0	3	2
Poultry	0	2	6
Sundries to farm	0	2	0
Attending and feeding pigs	0	4	0
Superintendence, education, and accounts	0	8	0
	7	4	2

Family.

	£	s.	d.
Attending dining-rooms	0	2	6
Steaming potatoes and other vegetables	0	2	0
Attendance on dormitories	0	1	8
Sewing and repairing beds	0	0	10
Washing clothes	0	5	0
Infant schoolmistress	0	2	6
Sundries to family	0	1	2
	0	15	8

Improvements.

	£	s.	d.
Carpenter's labour	0	10	8
Smith's ditto	0	9	0
Attending Slater	0	2	0
Storekeepers	0	16	0
Clerk of Accounts	0	6	8
	2	4	4
Total	£10	4	2

By reference to the ledger account of the
same week it appears that the consumption of
produce for the entire community (50 adults
and 17 children) was for that week as follows:—

	£	s.	d.
Potatoes, 243 stone	2	0	6
Milk, 202 quarts	0	16	10
Butter, 13¾ lb.	0	9	2
Mutton, 9½ lb.	0	3	2
Eggs, 32	0	0	8
Fuel	0	0	7½
Lodgings for single members and rent of			
Cottages for married ditto . .	0	10	0
	4	0	11½
Balance for clothing, etc., and for savings	6	3	2½
	£10	4	2

Results of the Experiment.

Thus it appears that the payments made for
labour during this week amounted to £10 4s. 2d.,
but owing to the superiority of the associative
arrangements over the isolated and competitive
system, the Association had a market on the
spot for a considerable proportion of their own
produce, and thus returned to the store for
articles consumed nearly one-half of the drafts

of the members, on account of labour, which
were on the same scale as the wages of the
neighbourhood (farm labourers, 8*d.* per day ;
ploughmen, having care of two horses, 10*d.* ;
herdsmen, 1*s.* ; blacksmiths, 1*s.* 4*d.* ; carpenters,
1*s.* 4*d.* ; women field labourers, 5*d.*). There
thus remained a sum in the hands of the fifty
labourers of £6 2*s.* 3½*d.* for clothing and other
expenditure and savings. Roughly speaking,
therefore, it will be seen from a comparison of
these statistics that the annual income of a
member of the Ralahine Association, which at
the ordinary rate of wages in the district would
only amount to a little over £10, was increased
to something like £16, or more than 50 per
cent., by the adoption of the co-operative principle. To these material advantages we must
add the great social and moral advantages
which arose from the arrangements of the
Associated Home. As already stated, there
was a common dining hall, 30 ft. by 15 ft., for
the accommodation of the single members, and
such of the married as preferred public to
private meals, thus saving the labour of cooking
in their cottages. Over these rooms were two
large dormitories for the girls and unmarried

women. One of the women had charge of
and kept the dormitories in order, and another
attended to the committee, lecture, and dining-
rooms. The washing and cooking being done
in proper places away from the dwellings, and
the children taken care of in the schools, the
married women were enabled to perform their
day's labour for the Society and to keep their
cottages clean and neat with very little labour.
As to the sanitary result of this mode of life, it
is a remarkable fact that during the entire existence of the Ralahine Association there was not
a single day's illness among the members,
although great numbers outside the community
suffered from fever, whilst the mortality from
cholera in Limerick and the neighbourhood was
very great. Much, however, as one is tempted
to linger on these various benefits, natural and
social, which were the direct result of the co-
operative principle, I must hasten the story to
a conclusion. The last act before the final
catastrophe is thus summed up by Mr. Craig :—

"The members were full of satisfaction with the present
and hopeful as to the future. The harvest was a splendid
one. The new (reclaimed) land of twenty acres had yielded
an ample return for their extra labour, which had been the

means of adding to our tillage land without increasing the rent. Six new dwellings had been erected by our own labour. It was also expected that the anticipated surplus would add twopence a day to the wages of labourers receiving 8*d*., and one penny to women's wages, being an addition of one-fourth in one case and one-fifth in the other. These advantages were in addition to those secured by wages, such as a second suit of Sunday clothes, while their children were clothed, fed, and well educated out of the common fund of the Society, and all of them had labour notes in reserve. Beyond these acquisitions and advantages the Association had, by their combined labour and care, produced and delivered to the landlord, as rent and interest, for the land, stock, buildings, and machinery, the following quantities of produce :—

	£
46,400 stones of wheat at 1*s*. 6*d*. per stone	. 480
3,840 stones of barley at 10*d*. per stone	. 160
480 stones of oats at 10*d*. per stone .	. 20
70 cwt. of beef at 40*s*. per cwt. . .	. 140
30 cwt. of pork at 40*s*. per cwt. . .	. 60
10 cwt. of butter at 80*s*. per cwt. . .	. 40
	£900

" These great results had been realised within three years at Ralahine, and others, with the right men, might follow our example. Leaders and organisers, sufficiently enlightened as to the principles involved in the new system or science of society, with all the higher 'resources of civilization,' could call into existence similar associations in a short time, and establish them in every county of Ireland, and what a wondrous change would be seen in the green isle of the ocean ! As it has been truly said, if our system had been allowed to continue, its example might have helped to make Ireland a paradise of peace."

The Final Catastrophe

Just at the moment, however, when the experiment had become successful beyond all expectation, all the high hopes for the future of the little community were suddenly dashed to the ground by the startling intelligence of their founder's absolute ruin and bankruptcy. Mr. Vandeleur, it appears, though a high-minded and benevolent man, was disastrously addicted to gambling. At his club in Dublin he indulged this passion to the extent of sacrificing to it everything he possessed in the world. His total ruin fell on the happy society of Co-operators at Ralahine with the effect of a thunderbolt. A distant relative of Mr. Vandeleur, a banker at Limerick, through some technical point in the law, took advantage of the President's position, in connection with the Society as a manufacturing association and a trading store, to obtain a fiat of bankruptcy against the estate. As to the co-operative members of the Association themselves—

" The world was not theirs, nor the world's law."

Mr. Gladstone's Land Bill, with its recognition of the justice of tenant-right, was still in the far future. The members were held to be common labourers, with no rights or claims for improvements, as all they had created and added to the estate belonged to the landlord and his creditors. The original " agreement " was treated by the lawyers as so much waste paper. In the eye of the law, I suppose, they were right. But it was robbery nevertheless ! The members had paid their rent, yet they were remorselessly evicted. They had no remedy. Ruin came upon them suddenly, and social co-operation at Ralahine was at an end.

Short-lived, however, as was this Irish experiment, it has put on record a valuable experience as to the possible results of the application of the co-operative principles to agriculture, some of the lessons of which, both by way of warning and encouragement, I shall endeavour to summarise in the next chapter.

CHAPTER VII.

LESSONS OF THE RALAHINE EXPERIMENT.

"If each to each be all he can,
A very God is man to man."

CÆCILIUS.

THERE are one or two questions, mainly, perhaps, in the tone of objection, which will, I think, naturally suggest themselves to any practical man who may have taken the trouble to follow the story of the remarkable success, both social and economical, of the Co-operative Farm established by Mr. Vandeleur at Ralahine.

The first and most obvious question will, I think, be this :—If, as you say, the adoption of the new principle of profit-sharing on the part of Labour with Capital on the Ralahine estate was found to work such wonders, not only with regard to the productive results of the farm, but also with regard to the social relations both of the labourers with one another and with their landlord, how does it come about that

during the fifty years that have elapsed since
then no Irish landlord has been either public-
spirited enough or far-seeing enough to follow
Mr. Vandeleur's lead?

Agricultural Conservatism.

I am afraid the answer to that question which
commends itself to my mind is not one which is
very complimentary to the Irish landlord. For
I cannot help feeling that the chief reason why
the "new system" should have been so little
productive of other experiments in the same
direction is to be found simply in the fact that
it was a *new* system that was proposed for imita-
tion. After all, perhaps, it is well to remember
that it is not only landlords, either in Ireland or
elsewhere, who dislike "new methods."* The

* The following extract from the evidence of Mr. Row-
landson (the Liberal tenant-farmer candidate for the North
Riding last year) before the Duke of Richmond's Commission will be interesting in this connection.

" 'If you consider that an objectionable system, why do
you not change it in regard to your own tenant, to whom
you are in the relation of a landlord?'—'I should be very
glad to do so as soon as a general system of that kind is
adopted.'

" 'If it is right and proper and fair to the people of the
neighbourhood, why should you wait until it is the general

fact is that most of us are by nature the child-
dren of custom—constitutional conservatives by
heredity. Liberalism, after all, the desire for
improvement and progress, is evidently a "state
of grace," and a state of grace moreover which
rural environment seems especially unsuited to
foster or encourage. Not, of course, that I would
for a moment seem to imply that every territorial
Conservative is quite so hopeless in this respect
as the old squire of whom George Eliot tells us,
whose whole political and economical creed was
summed up in the words " Whatever is, is bad,
and any change is likely to be worse ; " but at
the same time one cannot help feeling that
few things are so disturbing to the rural mind
as criticism of the old customs, and that, quite
apart from the question as to whether the
customs criticised are good or bad. The sort

system of the neighbourhood?'—'I suppose that if I adopted
a system like that in the neighbourhood, I should be something like a black sheep among the flock, but I should be
quite willing to adopt it if it were generally adopted throughout the neighbourhood.'

" ' But why is it necessary to adopt it generally in the
neighbourhood, and not in every individual case?'—'Your
Grace will be aware that there is always an aversion to a
person adopting a new principle, especially in a small and
simple case like mine.' "

of hesitation expressed in the lines of Clough is, I take it, a far too common attitude of mind in this respect—

> " Old things need not be therefore true,
> O brother men, nor yet the new ;
> Ah, still awhile the old thought retain,
> And yet consider it again."

If, however, I can only hope that the story I have endeavoured to relate would induce even one or two of my readers to "consider" this old experiment at Ralahine " again," I shall be amply satisfied.

There are two other questions of a practical character that are likely to occur.

I. It may be said, still mainly by way of objection, " Is this so-called 'new system' anything more than a device for transferring wealth from one body of men to another, from the landlord, that is to say, to the labourer, and thereby enriching the latter at the expense of the former, by the simple expedient of depriving him of a portion of his legitimate earnings to distribute it amongst his employés ? "

To this I would reply, Certainly not ! The principle of Participation in Profits is by no

means a mere philanthropic device for transferring the property of the rich to the poor,
but is, in fact, a method spontaneously capable
of realising additional profits, and thereby of
actually creating the fund which it proposes to
divide.

This point is important. Let me adduce one
or two reasons and facts in its support.

Stimulus of Profit-sharing Principle.

And, first, there is no more common assertion,
I think it will be generally acknowledged, on
the part of the farmers when discussing the
capabilities of the agricultural labourer, than
this—that the same man, in his employer's field
and in his own garden or allotment, presents
two surprisingly different standards of activity.
No doubt the assertion is true ; but the reason
is obvious. In the former case there is the
deadening certainty that no additional effort
will bring additional wage ; and in the second
case there is the enlivening hope that every
stroke of efficient labour will bring with it its
appropriate reward. Hence it is plain that
fixed wages tend to produce a minimum
standard of work, whereas the stimulus of

personal interest as inevitably tends to a maximum standard. Now the principle of profit-sharing adopted at Ralahine manifestly supplied such a stimulus. Take this fact in illustration of the argument. A certain visitor to Ralahine on one occasion happened to find one of the members of the Association at work *and alone*, under the following circumstances:—

The watercourse which supplied the power for the threshing machine, as it left the lake on the estate, passed under the old mail road from Limerick to Ennis, and near the tunnel the masonry had· given way and obstructed the flow of the stream. The visitor was surprised to find one of the members standing up to his middle in the water, repairing the wall, and entered into conversation with him to the following effect :—

Visitor. Are you working by yourself?

Member. Yes, sir.

V. Where is your steward ?

M. We have no steward.

V. Who sent you, then, to this kind of work ?

M. The committee.

V. What committee ?   Who are the committee ?

M Some of the members, sir

V What members do you mean?

M. The members of the New System—ploughmen and labourers.

The fact was, as another of the labourers once said to Mr. Finch, another visitor to the establishment, "We formerly had no interest either in doing a great deal of work, doing it well, or in suggesting improvements, as all the advantage and all the praise were given to a tyrannical taskmaster, for his attention and watchfulness. We were looked upon merely as machines, and his business was to keep us in motion; for this reason it took the time of three or four of us to watch him, and when he was fairly out of sight, you may depend we did not hurt ourselves with too much labour. But now that our interest and our duty are made to be the same we do not need any steward at all."

Here then is the answer to that first question. Improved work, spontaneously given, brings with it, in general, increased production, better quality, less waste and diminution in the cost of superintendence. This means, of course, enhanced profits. The principle of participation thus rests on a firm economical basis—viz., the creation by the more efficient labour called forth under its influence of new profits which do not accrue under remuneration by fixed wages only.

II. Can agricultural labourers be induced by the prospect which participation offers to put forth the sustained exertions necessary to secure its benefits?

The two last chapters have already in effect answered this question. I may add, however, this additional testimony from Mr. Craig :—

"At harvest-time the whole Society would voluntarily work longer than the time specified, and I have seen the whole body occasionally at these seasons act with such energy and accomplish such great results by their united exertions that each and all seemed as if fired by a wild, enthusiastic determination to achieve some glorious enterprise—and that, too, without any additional stimulant administered to them in the shape of any pecuniary reward."

Lessons of Encouragement.

So much then by way of reply to possible objections. As to the lessons of encouragement to be learnt from this experiment, they may, I think, be briefly summarised as follows :—

1. *Increased Stability of Relations between*

Capital and Labour.—The verdict of the Ralahine labourers themselves on this head was given in a testimonial to Mr. Craig at the close of the undertaking. They wrote :—

"We, the undersigned, have experienced for the last two years contentment, peace, and happiness under the arrangements introduced by Mr. Vandeleur and Mr. C. T. Craig. At the commencement we were opposed to the plans proposed by them ; but, on their introduction, we found our condition improved, our wants more regularly attended to, and our feelings towards each other were at once entirely changed from jealousy, hatred, and revenge, to confidence, friendship, and forbearance."

2. *The Beneficial Educati·nal Effects of Corporate Opinion.*—In illustration of this point let me give two brief quotations from Mr. Craig's book :—

" The weekly meetings had the happy result of interesting the members in the proceedings of the committee and in the success of the new system. The views of the committee recorded and read from the 'Suggestion Book' were discussed, and the practical value of certain methods of dealing with the land were also discussed, and formed an excellent basis for the educational training of the people and the formation of correct notions and higher phases of character than is possible with the usual isolated methods.

" There were at first two or three fellows inclined to be idle, and they were cured in the way wild elephants are tamed. The committee who knew their characters fixed their labour, and appointed one of these idlers to work between two others who were industrious—at digging, for instance ; he was obliged to keep up with them, or he became the subject of laughter and ridicule to the whole society. This was what no man could stand."

3. *The Preservation of Property and avoidance of Waste :—*

" During the winter of 1832, a hunted fox crossed the mill water-course near the rickyard, and took across the orchard, and over a 70-acre field of wheat in the highest tilth of any land upon the estate. The mounted huntsmen, —young squires, farmers, and tradesmen,—to keep well up with the hounds on the wheat-field, would have to pass through the farmyard, but they found that by a sudden and mutual impulse the large high gates of the farmyard had been locked against them by the ' new systemites.' Many of the huntsmen seemed perfectly astounded at the daring and ' impudence ' of these men. The incident shows that the new system had converted these once indifferent or careless servants into prudent conservators of the property under their care."

Again continues Mr. Craig :—

" Before the Society was established the labourers conceived their own interest opposed to that of their employers, and would attend to nothing beyond their appointments for the passing moments. . . . They conceived it to be their interest to encourage clandestinely the destruction of property, believing that it would create a greater demand for their labour. But after the Society commenced this order of things was reversed. A single potato was by many of them reluctantly wasted, for they found that the conservation of property was the saving of their own labour. Thus the same faculty of mind—self-interest—produced opposite results when surrounded by opposite controlling circumstances."

4. *The refining Power of the Principle of Social Sympathy :*—

"A report reached Ralahine that the crops of a poor widow, who had lost her husband by fever or cholera, would be lost in consequence of the death of him who had sown but was not there to reap, and the absence of means to pay for reapers. On the Sunday following all the young men of the community took their sickles and cheerfully travelled to the desolate home, cut the poor widow's wheat, and harvested her crop free of all cost. This benevolence was shown in other similar cases. Had the members been in an isolated position they could not have done this generous work of charity, and it serves to illustrate the refining and elevating tendency of the principle of social sympathy arising and co-extending with the humanising influence of the new system."

5. *Encouragement of thrift by the direct provision in the present of a lucrative investment for savings.*

6. *Consequent decrease of pauperism.*

7. *Increase of the true spirit of manly independence and self-respect, owing to the consciousness on the part of the participating workman that he is no longer regarded as a mere pro-*

*ductive machine, but as a human being having
aims and interests identical in kind with those
of his employer.*

8 *Direct advantages to the consuming public
in the increase of genuine work and upright
dealing, which is always the result of the corporate as opposed to the competitive spirit of
trade workmanship.*

Quotations in illustration of all these considerations might easily be given from Mr.
Craig's pages.

9. *The direct promotion of true religion, inasmuch as a system which takes for its root-
doctrine—" Society exists only for the sake of
the individuals who compose it, not merely to
further the accumulation of capital"—and for
its watchword, " Human progress and well-being
through self-sacrifice and association," cannot
but be doing something to realise on earth that
" Kingdom of God and His righteousness" which
Christ came to reveal.*

" Co-operation will teach men that God's
moral law is as irrevocable as His physical
law—that it is not the law of 'living by
getting,' whose motto is ' Every man for himself,' but that it is the law of 'living by

giving, whose motto is, 'Each for all, and all
for each,' and that enlightened self-interest can
be attained only by the path of self-sacrifice.
In a word, it will teach them that the Sermon
on the Mount is not Utopian, but that its
Divine command is true to the very letter,
'Give, and it shall be given unto you ; good
measure, pressed down, and shaken together,
and running over, shall men give into your
bosom.' The golden rule alone will bring the
golden age, and the Lord's Prayer will become
a reality when we see men uniting to give
generally to those around them the advantages
which they seek to secure personally for themselves. This simple rule will bring the Reign
of Righteousness ; daily bread will be secured
to the daily toiler ; God's Kingdom will come
as we His children learn to develop it, and
His Will be done on earth even as it is done
in Heaven, when we learn to obey it." *

* Lecture by Miss Mary Hart on "Ralahine."

CHAPTER VIII.

THE CO-OPERATIVE FARMS AT ASSINGTON AND BRAMTON-BRYANT.

"True wealth, as Carlyle somewhere says, consists in the abundance, not of the things which you possess, but of those which you take an interest in, and there are few English villages in which the raw material of such wealth does not abound for owners of the soil, capable of taking as much pride in the men as in the cattle that help to till it, and willing to learn from Mr. Gurdon one of the ways in which, with little trouble and no risk, the material may be suitably fashioned."—W. T. THORNTON : *On Labour*.

A YEAR or two before the commencement of the experiment at Ralahine, and fourteen years before the first enrolment of the Rochdale Pioneers, Mr. Gurdon, of Assington Hall, in Suffolk, came to conceive the idea of establishing a Co-operative Farm on his estate. During the fifty years which have elapsed since that time, the story of the Assington Co-operative Farms has often been told. Mr. Gurdon himself gave an account of them at the Social Science Congress at York in 1864, which is published in the "Transactions " of the Association for that year. Perhaps, however, the fullest and most interesting narrative is that given by the present Bishop of Manchester, in his Report, as assistant commissioner, on the Agricultural Commission in 1867. I cannot do better, I am sure, than quote his words in full :—

" In the year 1830, fourteen years, therefore, before the commencement of the enterprise of the Rochdale Pioneers, which has attained such gigantic proportions, the idea suggested itself to a Suffolk squire that he would attempt to apply, by way of experiment, the principle of co-operation or co-partnership to a farm. Selecting 60 acres of land of medium quality, furnished with a rough but not unsuitable homestead, he formed his little company of shareholders, all of them taken from the class of farm labourers, to which he gave the name of The Assington Co-operative Agricultural Society. The number of the original shareholders was 15, who put £3 a-piece into the concern, by way of subscribed capital ; the landlord, to give his bantling a chance of life, liberally advanced to the co-operators the sum of £400, without interest, on loan.* The society has grown and prospered. The occupation has been increased from 60 acres to 130 ; the number of shareholders has been enlarged from 15 to 21. The present value of the shares, as the bailiff told me, is 'all of £50.' All years have not been equally remunerative, but

---

The capital was below the ordinary estimate per acre ; but seems to have been sufficient to keep the land in good heart and cultivate it profitably.

there has not been one since the concern started without some little matter to divide. The company have repaid the landlord all the borrowed money, and all the stock and implements on the farm are now their own. The stock consists of six horses, four cows, 110 sheep, and from 30 to 40 pigs. The rent of the land is £200 a year, the company paying tithe, rates, and taxes. The farm is held on a fourteen years' lease, which is on the point of being renewed. The land is farmed on the four-course system of husbandry, and ordinarily employs five men and two or three boys. The members are not bound to work upon the farm, which, indeed, could not find employment for all; but it is understood, though there is no rule to the effect, that if a co-operator is out of work elsewhere, he has a claim to employment before any other man. When a co-operator works on the farm, he is paid wages at the usual rate ; and if he were not an efficient labourer, there would be no scruple about discharging him. The affairs of the concern are managed by a committee of four, but the practical direction of the farm rests with the bailiff, himself a co-operator, but employed as a servant of the company, and paid 1s. per week above the usual rate of day wages. Some of the members of the committee cannot read or write. Two fresh members are elected in rotation every year ; and though want of scholarship would not exclude him, yet if a man were not thought sufficiently intelligent for the business he would have to discharge, he would be refused when his turn came. All the voting is by ballot. No member is allowed more than one share ; only labourers of the parish are eligible for membership ; and if a man goes to live three miles away from the parish, he must dispose of his share. As long as he remains a member, he must, by the rules of the society, be a member also of the

Stoke and Melford Benefit Club. A member can sell his share, with the landlord's and committee's approval. When a fresh member is admitted, he pays £5 down, and the remainder of the current value of the shares by successive instalments. The landlord chose the original members, and claims to have the approval of new members; but he does not interfere with the company, as regards the cultivation of the land, more than he would with any other tenant. The premises are required to be kept in repair by the tenants, the landlord finding rough materials. They are to be insured in the amount of £500, and every twelve years the farm is revalued. A member, falling into difficulties, can have a loan advanced to him up to half the current value of his share; a privilege, however, I was informed, which has rarely been used. The annual profits are divided equally among the shareholders. Among the members are four widows, one of whom has four small children; they do what they can for themselves, and up to the present time have been able to maintain themselves by their work and the dividend on their shares, without the aid of parochial relief. Indeed, the guardians would disallow relief in the case of any person possessed of property of the amount represented by the value of a share, so that the scheme has a direct tendency to diminish pauperism.

"The first experiment apparently succeeded so well, that in 1854 Mr. Gurdon was tempted to try a second, and started the 'Assington Co-operative Agricultural Association.' The new concern began with 70 acres of land, and 36 members, each subscribing £3 10s. by way of capital. Again, the liberality of the landlord was taxed to supplement this inadequate amount of capital by a loan, without interest, of £400. The company has so far prospered that, though the times have been somewhat hard with them in consequence of the burden of this debt (which is now, however, wholly repaid), and the taking in and stocking a considerable accession of land, their present condition is as follows :—they now occupy 212 acres, at a yearly rent of £325, the company paying tithe, rates, and taxes, which amount to about £50 a year. The company is entirely out of debt ; the stock of the farm is valued at £1,200 ; the original £3 10s. shares would sell freely for £30. There has not yet been anything worth speaking of in the way of profits to divide ; and what has generally been distributed has been some article in kind, as a ton of coals, or something of the sort, to each shareholder ; but the members are satisfied with the state of things, and the prospects of the concern are bright in the future. All the members but six are of the class of farm labourers ; the six excepted ones are a miller, a blacksmith, a shoemaker, a wheelwright, and two carpenters. Female labour is only employed at weeding time, or for a job of stone-picking ; and at present there are only three boys working on the farm, one of ten, the second of fourteen, the third of sixteen years of age.

"The societies are not yet incorporated, but intend to be. The squire, I believe, has ceased for some years to be resident at Assington ; so that the success of the two experiments may fairly be set down, not to any sentimental fondling on his part, but to the sound principles on which they were based, and the prudent management by which they have been conducted. The only exceptional advantage which the societies have enjoyed, as they do not appear to be at all favoured in the matter of rent, was in the landlord's original loan, in both cases now paid off, to enable them to stock their farms.

" I paid a visit to Assington, to see the phenomena with

my own eyes. I gathered the information, which I have summarized in the preceding paragraphs, from Mr. Hedges, a large occupier in the parish, and churchwarden, and from the two bailiffs, John Crisell and John Marshall, upon the co-operative farms. Judging as well as I could judge from appearances, I have no hesitation in saying that the experiment has been an eminent success, and that it is an experiment well worth trying in other localities.*

"The only objections of any force taken to it were that of Mr. Hedges, that if the system became general, it would extinguish the tenant-farmer class ; and that of Mr. Maud, that the tenant-farmer class being extinguished, there would be a chasm in our social, and particularly in our parochial, system that it would be difficult to throw a bridge over. But these objections, though theoretically forcible, may be practically disregarded. It is not likely that the small-

---

* It is an experiment as it seems to me that many a clergyman might find it advantageous, both to himself and to his parish, to try upon his glebe. I read recently an interesting paper addressed to the Newbury Farmers' Club by Mr. F. W. Everett, in which the writer lamented, on economical and social grounds, the disappearance of small farms, meaning by " small farms " holdings between the size of 50 and 250 acres. He considered that many articles of daily consumption, requiring close personal attention, such as poultry and stock, were produced more successfully on small farms than on large ones. The co-operative system would encourage the reappearance of small farms, without the reappearance of a class that neither did themselves nor any one else much good, the class of small farmers.

farm system will ever become general, or the capitalist tenant farmer be displaced by a body of co-operators little if anything above the rank or intelligence of labouring men. No landlord would retrace the steps of the last half century, and break up his estate again into a number of small holdings. I think it very questionable if these co-operators would be able to manage a larger business than they are at present conducting. I am not at all the more assured of the performance and solidity of the great Rochdale enterprise because I am told that there is invested in it, in one form or another, a capital of several hundred thousand pounds. Concerns may become too unwieldy to be manageable, too gigantic to be safe. The very success of the Assington experiment appears to me to be due, in part, to the moderate limits within which it has been carried on.

"Mr. Maud, though thinking that the drawbacks of the system outweigh its benefits, enumerates among these latter some very considerable items. It attaches, he says, the labourer to his parish, in fact, to the soil. It counteracts the drain of which farmers so loudly complain; that is, of their best men into other employments supposed to be more remunerative. It is a decided help, Mr. Maud allows, to the labourer in a pecuniary point of view; and if widely adopted, would greatly diminish poor-rates—*i.e.*, pauperism; and, with pauperism, crime.

"To these admitted advantages may be added others. At the same time that the system displaces no labour, the co-operative farms employing no more hands than if they were occupied by a single tenant, it diffuses among a much larger number of the population an interest in the soil, and with that, an interest in the prosperity and stability of the country. In these revolutionary days, the tendency of the system is decidedly anti-revolutionary. The co-operator is a man who knows and feels that he has something to lose. And not only so, but the system increases that honest spirit of independence and self-respect which I am sure is as necessary in the lowest class as in the highest to rescue it from degradation. Mr. Maud says that it has not yet done much for education ; but I think it, infallibly, will do. It is hardly conceivable that a system which has such a direct tendency. to develop the sense of personal interest, should not at the same time develop a desire of knowledge, which may be called the correlative of the sense of personal interest. It is the poor drudge, to whom to-morrow is as to-day, without prospect and without hope, who is content to remain in his ignorance."

Unfortunately the societies have never been registered under the Provident Societies Act, and as they have never published an Annual Balance Sheet, reliable information as to their financial position has not been forthcoming. I have on several occasions made an effort to procure such information, and have failed. All that it was possible to discover was that of late years, at any rate, No. 1 Farm was continuing successfully to pay its way, while it was re-marked that No. 2 Farm was feeling the strain of the bad seasons of the last few years. The truth of that rumour has been made evident by the Letter of Appeal which has just been issued

by the Guild of Co-operators. The following
is the substance of that Letter:—

"At the Local Conference of Co-operative Societies in
the Colchester District, held at Harwich on the 2nd instant,
a Report on one of the well-known CO-OPERATIVE FARMS
AT ASSINGTON (Suffolk), known as the '*Severals* and *Knotts
Farm*,' was submitted; and after a long discussion, the
following Resolutions were unanimously adopted :—

"' Resolved, that the facts now stated show that the
"No. 2" Farm at Assington would probably regain its
former prosperity if sufficient capital were provided by the
Co-operative Societies, especially those in the neighbouring
Counties,—and if those Societies supplemented the custom
provided by the general market.'

"' Resolved, thereupon, that a Deputation be sent to the
Societies, inquiring *what help in Capital and Custom those
Societies would contribute if a new Society were now formed to
carry on the Farm in question.*

"' Resolved, that "the Guild of Co-operators," which has
for some time past inquired into all the circumstances of
the Farm, be entrusted with the duty of carrying the
preceding Resolutions into effect.'

"We are therefore instructed by the Guild Executive
Committee to bring this subject to the notice of Co-
operative Societies in Suffolk, Essex, and neighbouring
Counties, in order that their respective Committees may
consider the following facts, and decide *whether it is desirable that a Society should be formed to carry* on the 'Severals
and Knotts Farm,' and, if so, what capital and what custom
can be contributed by the existing Distributive Societies.

"The Farm in question, together with the other in its
immediate vicinity,—'the No. 1 Farm,' have been carried

on since the years 1829 and 1850 respectively by two Co-operative Associations of labourers, and have been repeatedly mentioned in the works of Economists and in the writings of social reformers generally. Their success in the hands of Agricultural Labourers, united in co-operation, has been referred to as a remarkable proof of what can be done in agricultural production so conducted.

"The two Associations not only paid off the whole of the capital advanced to them, including the value of the stock, but for many years paid a regular money dividend. Indeed, the profits were supposed to be so good as to indispose the members to admit new Associates. Their superior comfort and well-being were generally a subject of remark, and the present Vicar of Assington has borne testimony to the social and moral influence of these Associations on the inhabitants generally.

"In consequence, however (1) of the long series of disastrous years which have ruined hundreds of farmers; (2) in consequence of inadequate capital* for replenishing the farm with fresh stock, for making improvements, and for rearing cattle† and sheep, working a dairy, etc.; and (3) in consequence, perhaps, of defective management, recently, the No. 2 Farm has now to be wound up. There is, how-ever, enough to pay all debts, including the Landlord's Rent, and to leave a considerable surplus. It must also be borne in mind, that the other Farm continues to pay its way, and is in good condition.‡

---

* No reserve fund was formed, and profits were all paid away.

† Thus manure will be provided by the working of the farm, instead of being imported, as at present.

‡ I should myself venture to add, "And (4), perhaps the

" Now the question has arisen whether it is not desirable that *a new Society should be constituted to carry on this Farm.* Much has been said as to the importance of Co-operation being applied to Agriculture, in order to raise the condition of labourers and to enable distributive Societies to obtain Farm produce direct from the producer. Indeed, Co-operators have to justify their claims to be able to apply Co-operation successfully to this form of Industry as well as to all others.

" We submit the facts connected with the Farm in question.

" It comprises 223 acres of land, consisting of a rather heavy soil, for the most part, some of it well adapted for cereals and root-crops, and some for pasturage. It is stated that mutton can well be supplied, vegetables and dairy produce, as well as the crops hitherto raised.

" The Farm is four miles from the Railway Station at *Bures.*

" The rent is £268, being 24s. an acre (which is 8s. an acre less than what was formerly paid).

" The Landlord would be willing to grant to a Registered Society a lease of seven or fourteen years.

" He will impose no restrictions and conditions as to the mode of farming ; an obviously important advantage.

" He will make no reservation as to rabbits, giving unestricted permission for their destruction (excepting during ' closed time,' when the birds are nesting). This is an advantage not possessed in the past, when loss was very great from this cause.

---

most important consideration of all, in consequence of there being no regulation that the labourers on the farm should also be shareholders of the Association."

9

"The farm is fairly well stocked, even at present; and here is a supply of agricultural machinery, houses, stables, etc.

"It is considered that, with a capital of £2,500, the farm could be placed in excellent condition, and its prosperity assured, by extending and improving production, by increasing the number of cows and horses, by rearing mutton, by establishing a dairy, and by securing first-rate management.

"If a new Society were formed, a large number of the present farmers would become members and give their long practical experience; whereas, on the other hand, the representatives of existing distributive Societies would bring a wider experience, better business habits, and fresh energy. The open market would be supplemented, to some extent, by the demands of those Societies for vegetables and grain, for milk, butter, eggs, fowl, mutton, bacon, and other articles.

"Of course the question of the extent to which Societies can supply themselves from the Farm must depend on their distance and the cost of carriage. They might at once set on foot inquiries as to this point, and place themselves in correspondence with the secretary of the No. 2 Farm Association, Mr. Pollard, who attended the recent conference at Harwich, and gave some useful information.

"The general question of forming a new Society will have to be looked at from several aspects :

1. "The importance to the progress of the movement generally, that co-operative Agriculture should gradually be tried under conditions as favourable as possible, and the importance of avoiding the discouragement which would result from total abandonment of the Assington Farm.

2. "The question whether this particular Farm is likely

from its character, its position, and the special circumstances, to give fair guarantees for success as an investment.

3. " What are the advantages which the existing Societies
would derive from supplying themselves with produce in
this case, direct from the producers, who would be identified
with them as closely as possible.

" A few details may be added, as enabling the Societies
to determine what course they will take.

" Mr. Pollard believes that the present members would
take up shares (which would probably be of the value of
£1 each) to the extent of £500; which would leave only
£2,000 to be raised by Societies and individual members
in adjoining and other Counties.

" The present Association proposes to wind up by selling
everything on the farm by auction and paying all liabilities
from the proceeds. The new Society, if formed, could
purchase at the sale such articles as were in good condition;
obtaining at other sales (at Michaelmas) further requirements.

" In addition to the rent, tithes and rates of all kinds
amount to £52 per annum, which sum, with the rent,
makes a total of £318. It should be observed, however,
that the present rent is not high, and is *eight shillings* an
acre less than what the members paid in their prosperous
years.

" The Landlord is willing to treat with a new Society,
and Mr. E. V. Neale has prepared Rules for one. Societies
which are situated too far to avail themselves of the farm
produce might, nevertheless, be willing to take shares in
the Society formed to carry on the undertaking.

" The Guild Council heartily recommend the proposal to
form a Society for carrying on this Farm to the best consideration of Co-operative Societies and their members."

A Co-operative Farm in Herefordshire.

The account of another unsuccessful attempt at direct Co-operative Farming in England I may also perhaps conveniently give here. It is that of the Brampton-Bryant Farm, established by Mr. Walter Morrison in 1873. The following is an extract from a letter which Mr. Morrison was good enough to write to me on the subject :—

"I was led to make the experiment somewhat by accident. I heard, in 1872, of the formation in North Herefordshire of an Agricultural Labourers' Union : this was long before Mr. Arch was heard of, and I may observe in passing, that it was very successful. Its funds were small, but they were carefully and honestly administered. There were no extravagant expenses of management, the salaries were ridiculously small, and until the health of the Secretary, Thomas Strange, the soul of the Union, broke down from sheer hard work, it flourished. Wages rose greatly,—the funds were chiefly used in removing surplus labourers to the North,—and though the farmers naturally dislike the movement, there was no strike, and it was not discredited by the spiteful abuse of squires, parsons, and farmers, which have disgraced other organizations. I learned that its leaders were men who had accumulated sums up to the amount of £100 in the Savings Banks, in a district where the wages were nine and ten shillings a week. In July of that year I paid a visit to the Co-operative Farms at Assington, with the Hon. and Rev. J. W. Leigh, who had been starting a Co-operative Store in his village. He told me of a small farm which was on sale near Leintwardine, which was the head-quarters of the Herefordshire Labourers' Union. The farm was half arable and half grass, contained 148 acres, and was let for £140 a year. I went down to see the farm, and to form some idea of the capabilities of the leading men of the Union. I was very much pleased with them, and so having purchased the farm, I sent over three of the Union Leaders to spend a few days at Assington. They came back in a state of great enthusiasm about Co-operative Farming. The result was, the formation of a Society, whose rules I forward with this. In March 1873, they entered upon the occupation of the farm at the old rent. There were then twenty-seven members, with a capital of £510. In October 1873, the members were twenty-nine in number, with a capital of £662. Of this sum £467 were contributed by twenty labourers or artizans living near, and £195 by nine friends of mine, who were interested in the experiment. When we wound up the Society in 1879, the farm having been given up on March 25th, 1879, the share capital was £808, and the number of members, I think, thirty-one. At the time of our start, all stock and implements were at a very high price, while they were at their lowest price when we had to sell them off. This must be allowed for in estimating the financial result; but no dividend was ever earned during the tenancy, though at the end of the first year the Committee credited each member with five per cent. interest, without, however, paying out any cash on account of the dividend aforesaid, and I found this supposititious dividend entered in the books, so there it remained; and when we wound up and sold off our assets, we found that with the help of rather a liberal estimate of what was due from me on account of tenant right, which I gave to them, as I have for many years to my other tenants, there was just enough to pay nineteen shillings in the pound, which, with the shilling in the pound improperly credited to the capital as mentioned above, really was a distribution of twenty shillings in the pound."

The causes of the failure Mr. Morrison considers to have been moral, not economical. The labourers were unsuccessful in finding the right man as manager. They tried two. The first was not honest, and drank. The second was perfectly honest, but lacked energy and administrative power. Altogether, it is evident that the labourers at Brampton-Bryant were not quite ripe for the movement.

CHAPTER IX.

AN EXPERIMENT IN CUMBERLAND.

"Never was landlord more sagacious, inventive, genial, or liberal—or changeable, not in his general purpose, but in his methods. Had he been less paternal, and taught his people the art of self-help, he had been a great benefactor."
—G. J. HOLYOAKE : *History of Co-operation.*

IF the story of the Co-operative Farming experiment at Ralahine in Ireland, related in the chapters v.—vii. might rightly enough be called "a romance in facts and figures," I fear some readers at any rate may be inclined to think that the experiment of which I desire now to give a brief account will rather merit the title of burlesque comedy or even screaming farce. Certainly the book in which the story is told is one of the queerest books about farming on record.

The title of the work "Ten Years of Gentleman Farming at Blennerhasset, with Co-operative Objects," gives no fair idea of its contents. For certainly there never was before, and

probably never will be again, either such gentleman farming or such co-operation.   The account
of the origin of the experiment given by Mr.
Lawson—a brother of Sir Wilfrid, the wit of
teetotalism—is itself characterised by a degree
of moral candour, which is certainly as refreshing
as it is uncommon.   " Trained as a shooter of
animals," he says, " a hunter of Cumberland
beasts with hounds, and a trapper of vermin, I
found myself in the spring of 1861, in my twenty-
fifth year, without an occupation, without many
acquaintances, except among the poor, whom
I had learnt to despise because they spoke bad
grammar and took their coats off to work ; and
without the reputation of having been successful
in any undertaking except that of the mastership
and huntsmanship of my brother's foxhounds."
Riding up to London at this time, he somehow
hears of Mr. Mechi's celebrated farm at Tiptree,
visits it, becomes enthusiastic for the Tiptree
system,   and,   after   endeavouring   in   vain   to
impress on his father and his practical men the
many and great advantages of Alderman Mechi's
system of farming over the old jog-trot one,
accepts the offer of one of his father's vacant
farms to experiment upon as he chose.   " Had

I been differently circumstanced," he says, "my ignorance of farming might have seemed a great objection ; but it seemed to me then to be of the never-go-into-the-water-till-you-know-how-to-swim kind ; besides, any one could carry on what he understood, while it would require some cleverness to carry on what one did not understand."

Zeal without Knowledge

Accordingly he diligently set about testing the quality of the soil ; spent several weeks in travelling for agricultural information ; engaged his father's coachman as head man—not then having appreciated the shrewd advice of Mr. Stephen, in his "Book of the Farm," that while honesty is an important qualification for a shepherd, knowledge of a shepherd's business is even more so ;—bought a steam plough at a cost of £825, and ten tons of low-priced guano from a cheap dealer in the neighbourhood, and then finally fell in with an intelligent agricultural engineer, who told him he had made three great mistakes already—the steam plough, the cheap guano, and, worst of all, the coachman-farmer. "How profusely I laid out money—

pulling down miles of old fences, making thousands of yards of good new roads, draining the
land five feet deep and ten yards apart, and
taking thousands of tons of stones out of the
ground ; how during several years I bought and
fed animals and sold them at a loss ; how I
deceived myself and was deceived by others in
various ways—can be sufficiently well imagined
without being described." However, the whole
experiment in the minutest detail of failure or
success is most carefully and elaborately described in the book, from its commencement, in
May 1861, down to the moment when after his
farm buildings had been burnt down, and it
became finally clear to him that his "farming
was very far from being remunerative, or from
giving prospect of becoming so," he sold his
farm to his brother in August 1871.

It must not be supposed, however, that even
under its business aspect the story is without
its value even to practical agriculturists.   The
chapters on "Farming Losses," "Manufacturing Profits and Losses," "Shop-keeping and
other Ventures," "Varieties of Farm Cropping,"
"Field and Crop Balance Sheets," "Manures
and their Values," "Farm Labour and its Cost,"

"Experiments on Grass Manuring," and "Field Experiments on Potatoes," are all full of characteristic and suggestive information which cannot fail to be instructive.

It is with the social aspect of the story, however, that I am more especially concerned. From the first Mr. Lawson had determined that the labourers on his farm should partake directly in the profits. " In going about to get information, I had found one great difficulty common to good farmers (as well as to the other sort), namely, the difficulty of dealing with the labourer. . . . So I determined that my farm should be a co-operative one. . . . Co-operation began gradually to take the leading place in my mind, and soon became the chief object in my life ; so that I did not so much intend co-operation to serve the purpose of farming, as farming to serve the purpose of co-operation."

Accordingly he called his labourers together and explained to them the meaning of Co-operation, told them something about what had been done by Mr. Gurdon, in Suffolk, who as long ago as 1831 had let a farm to thirty labourers, which had been ever since carried on successfully on Co-operative principles, and finally gave them the opportunity of deciding for themselves on the question,—" Is it desirable that the workers on the farm should be direct partakers of its proceeds?" by taking their vote by ballot.

" Our voting urns were two bottles : one was ticketed with the word ' Co-operation,' and the other bore the inscription, ' Every man for himself.' What, then, was the result with these eleven people? Actually ten of them voted for ' Every man for himself,' and only one put into the Co-operative bottle!" After twice again trying the experiment of a vote, to which not the labourers interested only, but all the inhabitants of the village were now invited, Mr. Lawson, at last, in 1866, " offered Co-operation to all comers," in the shape of one-tenth of the profits for the workers. " I felt," he said, " that direct participation by the labourer in the profits of the farm would be an improvement upon the existing system of paying him by regular wages only, and I kept urging its desirability, in different ways, to various people, for several years. I saw, ultimately, that the expediency of measures was

not always to be judged of by the number of
people voting for or against them, and that if
I approved of the principle of *Partnership of
Industry*, it was for me to introduce it, and let
it find its own value as the time went on."

In the *Co-operator* newspaper, in May 1866,
Mr. Lawson inserted the following letter :—

"LABOURERS SHARING ONE-TENTH OF PROFITS.

"TITHE FOR THE TOILERS.

" BRAYTON, *Carlisle, May* 13*th*.

" DEAR SIR,—I enclose a card which I have circulated
among my workers. I have offered them one-tenth of my
profits, and thus hope to make their labour interested. I
began last week to issue tickets with wages ; these tickets
are to be given in on January 1st, 1867, when one-tenth of
the profits will be divided on them.

" I find that if, in paying, the paymaster forgets the
tickets, he is soon reminded by the workers. I cannot yet
see any improvement, but hope that the end of the year
will show some.

" I remain, yours truly,
" WILLIAM LAWSON."

The ticket was as follows :—

"CO-OPERATIVE LABOUR.

"TO MY WORKERS.

" I give tickets with wages, that you may obtain two

shillings out of every pound of profit, and thus have a direct money interest in the success of the establishment.

"WILLIAM LAWSON."

Unfortunately, however, owing chiefly to mismanagement in cattle, sheep, etc., there was heavy loss on this year's operations, and consequently no dividend to the labourers.

In 1868, Mr. Lawson gave up the bonus on labour idea, and offered to the public in general all profits on his capital (which was about £54,000) over 2½ per cent. (the only interest asked). These public profits, as they were called, were not to be divided, but spent upon public uses ; and a free school, reading room and library, a village hospital, a Christmas festival, etc., engrossed about £150 of them during the year ; so that after Mr. Lawson's 2½ per cent. (about £1,330) and this £150 were deducted, about £30 was left, and this, in accordance with the resolution of two public meetings, and with Mr. Lawson's assent, was divided as a bonus upon labour. Time, not wages, was the basis of division, so as to give an equal share to the female workers. This bonus gave about 3½d. per week, or 10s. for the nine months to those on full time. In

1869 Mr. Lawson again made the same offer,
but the profit again fell short of the 2½ per cent.

The Great Bonus Year.

The year 1870, however, was more fortunate, and is still recollected at Blennerhasset
as " The great bonus year." The following
notice had been printed on cards distributed
among the labourers at the beginning of the
year :—

"To my Workers.

" I shall give as bonus to ordinary time-workers in proportion to time worked (exclusive of extra and overtime)
one quarter of this year's declared income arising from
my present capital, clear of all current expenses for public
good ; but should such income exceed £1,000, I shall give
as bonus half its excess over £500.

"WILLIAM LAWSON."

The balance sheet for that year of the whole
establishment showed a balance on the right
side of £1,715 4s., and it must be borne in
mind that no rent for the farm or garden, nor
interest on the capital of the establishment, was
charged. Some departments of the establishment yielded a profit, and others a loss, that
year ; but the net result, in figures, was a gain
of the above-mentioned £1,715 4s.

Of this gain £546 4*s.*. 7½*d.* was awarded according to the above-mentioned announcement to the workers, as bonus on time, affording 4*s.* 2¾*d.* per week to each person—man, woman, or child—or £10 19*s.* 11*d.* to every full-time worker during the year.

After the distribution of this large bonus it was naturally expected that the business would continue improving, and that 1871 would show a good profit too.   Much interest was therefore taken in the " manifesto " to be issued for 1871, and more public discussion and advice was bestowed upon it than on any previous one. The following is a copy of it as finally agreed upon :—

THE   BLENNERHASSET   CO-OPERATIVE   ESTABLISHMENT.

CAPITAL, £32,780 4*s.* 8½*d.*

Approximate Investment of Funds, November 1st, 1870.

Assets.

	£	*s.*	*d.*
415a. 3r. 20p. of farming and garden land, farm buildings, machinery, and cottage property .	30,900	o	o
Shops and house property at Newcastle	600	o	o
Farm and garden stock, crops, implements	5,554	1	1½

Assets (continued).

	£	s.	d.
Two steam engines and plough tackle	1,300	0	0
Manures on land	676	15	0
Sundry other property, Blenner- hasset	751	2	2½
American Investments . .	8,447	2	6
Invested at 4 per cent. . .	2,639	0	0
Invested in various Co-operative Societies	42	4	6
Book debts	154	0	4
In the bank	478	2	10
Cash in hand	29	10	0
	51,571	18	6

Liabilities.

	£	s.	d.
Mortgage on 393a. 29p. . .	18,000	0	0
Book debts	245	9	2¼
Bonus to workers for the year end- ing Nov. 1, 1870 . . .	546	4	7¼
	18,791	13	9¾
Balance net capital . . .	32,780	4	8¼
	51,571	18	6

NOTICE.—Of the declared income arising from the above-mentioned capital, for the year ending Nov. 1st, 1871, I shall devote one-third to expenditure for the public good, one-third to the payment of bonus to my time workers, and one-third to my own use—the income to be declared and

10

the bonus paid before the end of 1871.   N.B.—The bonus
is not payable to delinquents, nor is it transferable.
January 2nd, 1871.   WILLIAM LAWSON.

End of the Experiment.

Once more, however, the profits of the year
were *nil*, and no bonus to labour was possible.
In the same year Mr. Lawson sold the farm,
and the experiment came to an end.

It must be plain, I think, to any unbiassed
reader, that no conclusions either for or against
co-operative farming can really be drawn from
this experiment.   Evidently there was no
sufficient method in Mr. Lawson's co-operation.
He felt—what every Christian man should feel
—that his property was a great social trust ; he
was extraordinarily benevolent and sincere,
superbly indifferent to Mrs. Grundy, but he was
also wanting in steadiness of will and purpose.

The Defects of Mr. Lawson's Co-operation.

The general result of his experiment is well
summed up by Mr. Glassbrook, who acted as
bailiff to the farm between February 1868
and the sale of the estate :—" Mr. Lawson's
views on co-operation, I think, are good and
sound. and he has taken great interest in this

movement. He would have done striking good had he fully carried out the branches entered upon ; but as soon as any new scheme was got into working order it was laid aside ; and in my opinion this was the main cause of Mr. Lawson's non-success. Mr. Lawson co-operated with his workers very successfully, all his offers to them being highly appreciated and well wrought for ; and the workers were just beginning to have full confidence in their employer when the establishment was broken up. The class of workers on the farm was quite a superior body, and well worthy of co-operating with." A similar testimony is borne by Mr. Holyoake in his " History of Co-operation."

Though Mr. Lawson spent £30,000 on this experiment, " it could hardly be said to be lost, since at any point of his many experiments he might have made money had he so minded. But he proceeded on the plan of a man who built one-storied houses, and as soon as he found that they let at a paying rent, pulled them down and built two-storied houses, and when he found that they answered he de-molished them and put up houses of three stories, and no sooner were they profitably occupied than he turned out the inhabitants and pulled them down. What he lost was by the rapidity of his changes, for he had sagacity as great as the generosity of his intentions."

CHAPTER X.

" As wine and oil are imported to us from abroad, so must ripe understanding and many civil virtues be imported into our minds from foreign writings ; we shall else miscarry still, and come short in the attempts of any great enterprise."
—MILTON.

AMONG the many eccentric epitaphs that are to be met within churchyard annals, I should suppose that none are more singular than one which is to be found in the graveyard of a little church in the Mecklenburg Highlands. It is the inscription which has been graven, in accordance with his own express desire, on the tomb of a well-known German economist, J. H. von Thuenen. The epitaph is as follows :—

$$W = \sqrt{AP.}$$

By this mathematical formula the economist desired to express the principle that the natural wages of labour, represented by the symbol W, would, in any State uninfluenced by foreign

competition, be an exact mean between the cost
of the labourer's subsistence, represented by A,
and the value of his production, represented by
P. While accepting the dictum of Adam Smith,
and the older political economists, that " the
produce of labour constitutes the natural
recompense or wages of labour," Von Thuenen,
in following out the proportion of recompense
due to labour, when subjected to the combined
influence of capital, rent, and superintendence,
contended that the payment of labour reached
only its true level when placed on an equality
with capital. If the rate of wages sank to the
starvation point, owing to the action of competition, could that, he asked, be called the
natural rate of wages? " The man," he contended, " who has passed his life in honest and
laborious activity till his old age, ought neither
to depend on the favour of his children nor on
the community. An independent, care-free,
and easy old age is the natural recompense for
the incessant exertion of his days of health and
strength."

Industrial Partnership.

It was with the view of putting his principle

into practice that Von Thuenen, in the year
1848, established a system of industrial partnership with the labourers on his own estate of
Tellau, the continued success of which up to
the present time furnishes us with one of the
most instructive lessons to be drawn from the
history of co-operative enterprise. In the *Co-
operator* newspaper of October 31st, 1868, Dr.
Brentano, the well-known author of " English
Guilds and Trades Unions," gave the following
account of Von Thuenen's experiment:—

"Long before the year 1848, it was Thuenen's wish to
make an arrangement by which his labourers might participate in the profits of his estate ; but as it is not general in
Mecklenburg for proprietors to have such kindly feelings
towards their labourers, Thuenen dared not do it. The
movement in 1848, however, obliged every one of the
landed proprietors in Mecklenburg to make concessions to
their labourers ; and though there was no movement on
Thuenen's estate—for he had always lived on such terms
with his labourers that there were none discontented—he
thought the moment propitious for the purpose of carrying
into effect what he had so long desired. The arrangements
he made to give his workmen a share in the profits were
as follows :—
"If, after the deduction of certain expenses, the revenue
of the estate should be more than 5,500 thalers, to every
one of his labourers ½ per cent. of this surplus produce
should be credited. The average revenue of Tellau during

:he fourteen years from 1833 to 1847 had been 7,500 thalers
(£1,125). Thuenen calculated that the share of each
labourer would be 10 thalers (30s.) annually, if the revenue
of the estate should remain the same ; and he pointed out
that if this revenue, in consequence of a better cultivation
of the soil, should be increased every year by 1,000
thalers (£150), the share of the labourer would not be
increased in the proportion of 75 : 85, but of 10 : 15. The
interest of the labourer, he thought by this arrangement,
would be most intimately connected with the increase of
production. The number of labourers to whom he gave
a share in the profits was twenty-one, and included amongst
them is even the man who watches at night."

Objects of Von Thuenen's Scheme.—" The end at which he
aimed by this arrangement he describes as follows :—

" 1. That the inhabitants of the village might immediately participate in the welfare, and also in the losses, of
the proprietor of the estate, and might form together with
him almost a family.

" 2. That the labourers might enjoy an income increasing continually and higher every year by the interest of their
capital.

"3. That before all, the labourer might be sure of a
happy old age, free from the cares of life ; that after having
spent his strong manhood in toil and activity, he might not
starve in his latter years, when health and strength had
departed, that he might not live on the charity of others,
or as a burden to his children, but even be enabled to
bequeath something to them."

A Savings Bank.—" He erected a Savings Bank for this
purpose. At Christmas the labourers' share in the year's
revenue of the estate is to be inscribed in their savings
book. For every thaler inscribed the proprietor of the

estate has to pay one groschen interest, which is about four and one-sixth per cent., and at Christmas the interest is to be paid. The capital inscribed in the books of the savings bank cannot be withdrawn until the proprietor of the share has attained his sixtieth year. As soon as he has attained this age, the capital shall be at his free disposal; if he should die previously, his heirs are to get the amount standing to his credit."

Successful Results.

In a letter to Professor Boehmert, dated 25th May, 1877, the present proprietor of the estate, the grandson of the original founder, states that "the experiment has realised all the requirements proposed and expected by his grandfather. It has attached the labourers to his estate. It has secured for them an old age free from care. It has diminished the poor rate. It has established better and more harmonious relations between proprietor and labourers. Some of the labourers had more than £75 to their credit in the savings bank in less than twenty years after the system was introduced. Up till 1876, the labourer's share in the profits had only three times fallen below the £1 10s. which Thuenen calculated beforehand it ought to reach, if the average revenue of the estate remained what it had previously been.

Those three years were 1849, when it was £1 8*s.*, and two very exceptional years, 1866, when it was only 4*s.*, and 1876, when it was only 15*s.* In 1863 it was £7 10*s.*, and in 1864 £7 15*s.*, and it has averaged about £4 a year.

Herr Berthold Wöbling, in a paper quoted by Böhmert, says of these experiments :—

"These earnings have a special source of their own, viz., enhanced production due to the industry and care of the labourers. Every practical farmer knows how imperfectly agricultural work is done by hirelings of all sorts, and how little what goes by the name of good superintendence is able actually to effect in securing good execution of work. The full effect of any work is brought about, not merely by intensified exertion of muscular force, but also by zeal and alertness of mind. Such an application of bodily and mental forces is only to be obtained from one whose entire interests are engaged. In fact, new springs of production are thus opened, and it is this which gives to the system its high agricultural importance. The labourer finds that his increased incomings are relatively speaking more easily earned than under fixed wages, because they include payment for carefulness as well as for efforts of brute force. A reciprocal influence on the habits of the labourers will also not fail to show itself. If they perceive that a successful result depends not merely on muscular exertion, but also on sustained orderliness and attention, they will find it more and more their interest to practise these virtues.

"The proprietor derives, independently of the pecuniary result, many advantages from the half-profit system. He has perfectly trustworthy labourers, and each piece of work is taken in hand at the proper moment. He is no longer obliged to urge and drive, while fretting internally at the many instances of neglect which he is powerless to prevent. When his back is turned, he knows that his business is as well attended as if he were directing it himself. He can dispense with all intermediaries, as no formal overseeing is required. Nevertheless the position of the managing head has grown in importance. He must show more than was formerly necessary that his management is sound, and that with regard to every department of his business he is firm in the saddle, for he now has a responsibility towards his associated labourers. He is more than ever bound to set them an example of diligence, economy, and other virtues, on the exercise of which the success of the whole undertaking depends. In short, the system demands a thoroughly competent man."

Herr Jahnke's Scheme.

In some respects perhaps even a more interesting experiment is that of Herr Jahnke, a proprietor of Bredow, in Brandenburg.

" Struck with the distress of the agricultural labourers in the years 1871 and 1872, he resolved to devise a plan by which each of his workmen should enjoy a *menschen würdiges Dasein* (a life worthy of a man), and the way should be prepared for their ultimate independence. In 1872 he entered into a contract with five labouring families to work his farm for five years at half-profits. A regular deed of co-partnery was drawn up. The proprietor hands over the land, the steading, and the stock to the Company at a valuation, and agrees to leave all his capital, amounting to £750, as a burden on the business at 5 per cent. interest. He reserves to himself, however, his dwelling house, garden, peat and woodhouse, fruit trees and vineries, the unrestricted use of the well, and liberty to take peat, wood, and game. For other things he makes a special bargain with them. . . Herr Jahnke is manager ; buys, sells, keeps accounts, and gets £45 a year for doing so. . . The labourers receive a fine, healthy, and commodious dwelling-house each, a bit of garden ground, free peat and wood for their private use, and in money wages the five collectively are to receive £2 12s. 6d. a week in summer and £2 5s. in winter ; or in other words about £25 a year each. . . . An annual balance sheet is to be made up on the 10th April every year, and the profits, after deducting expenses aforesaid, are to be equally divided between the proprietor on the one hand, and the five labourers on the other. . . .

As concerns the pecuniary results of the enterprise, the net profits divided in the year 1872-3 were £528 ; in 1873-4 they were £488 ; and in 1874-5 they were £549 ; making an average over the three years of £521. Of this the proprietor received half, £260, and each of the five labourers a tenth, £52. For the work of himself and his wife each labourer had thus, in addition to free house, garden, and fuel, £25 a year in weekly wages, and £52 a year in annual profits, or £77 in all.

Now, according to Von der Goetz's Government Report in 1875, the ordinary income of agricultural labourers in Brandenburg was then £32 a year, and the highest paid agricultural labour in all Germany amounted to no more than £33 a year. It is clear, therefore, that the system proved very advantageous to the labourers, and Herr Jahnke says that it also proved advantageous to the proprietor. He was never in want of labour, as he formerly used to be, and the produce of the farm was considerably increased. The work was better done than it was before, and was much more skilfully arranged. The men needed no super-intendence. If one of the five thought to scamp work, the other four remonstrated and kept him to his duty. Jahnke says that though in one sense he paid more for labour than he did before, in another sense he paid less, for the labour was more productive and was attended with less waste and destruction.

A Lame and Impotent Conclusion.

In 1877 the contract expired, and Herr Jahnke did not renew it, but sold his estate, which had been valued five years before at £4,500, for £5,700. *The experiment had been an entire success,* he says, *but he had made so many enemies by it, especially among the class of large landed proprietors, that he resolved to give it up and sell his estate.*

It is indeed a mean and impotent conclusion! However, we may have some hope, I trust, that on the English side of the German Ocean at any rate, if only the profit-sharing principle can be shown to be commercially advantageous to both employer and employed, the system will not fail, if once attempted, from mere want of moral courage, as in the case of Herr Jahnke.

Social cowardice is not an unknown factor, of course, in English country life, and may perhaps even for a time throw obstacles in the way of the adoption of the co-partnery system in agriculture; and yet, as I am glad to think, not every English squire or landlord is afraid to face the patriarch Job's question—" Do I fear a great multitude, or does the contempt of families terrify me ?" (Job xxxi. 34). For already, as I write these words, proof comes to me that in one English county, at any rate, the system is not to go without fair trial. A Warwickshire landlord, whose name I must withhold for the present, sends to me the following notice which he has just circulated among the labourers on his estate :—

" The Manor Farm.

" I agree to distribute among all labourers working on the Manor Farm (under the conditions specified below) 60 per cent. of the net profits of the farm. These net profits to be calculated after deducting from the gross profits £300 a year for rent, 4 per cent. on £4,000 invested on the farm, and the bailiff's percentage on dairy and poultry sales.

" This money will be divided at the end of every year in proportion to the total amount of wages earned by each labourer during the year.

" No labourer shall be entitled to receive any share who has not been working on the farm regularly for twelve months at the time of distribution. If he is absent from his work for a single day without permission from the bailiff, he will forfeit his share.

" One-half of the money due to each labourer I shall invest for him at the Post Office Savings' Bank, and if he withdraws it without my permission, he will forfeit his right to any future share.

" This agreement will come into force from November 1, 1882, so that the first distribution of profits (if there are any) will take place on November 1, 1883.

" I reserve the right of cancelling this agreement at a year's notice."

NOTE TO CHAPTER X.

Mr. Bolton King's Co-operative Farming in Warwickshire.

At the close of my tenth chapter I spoke of the experiment in the co-partnery system in agriculture which a Warwickshire landlord was then about to make. That gentleman was my friend Mr. Bolton King of Gaydon, who, in the year 1883, organised the Radbourne Manor Farming Association. After a fair trial of seven years he has been obliged, owing to the heavy and continuous losses incurred, to discontinue the experiment. He writes to me under date August 20, 1890.

" I have had to decide to give up the farms. I found I could not stand the continued loss, as there does not seem to be a sufficient prospect of success to make it worth while losing more money on them. I am sending an account I published in a local paper, which gives what I feel to be the causes of failure. It is a disappointment of course, but every experiment I suppose has its value. I think co-operative farming will have to come about more gradually through allotments. Many villages about here have their allotment societies. These are of course a germ of co-operation, and they are already beginning on a small scale to make common purchases. This is sure to develop, especially as the men get capital accumulated in the village store."

The following is the extract from the *Leamington Chronicle* to which Mr. King alludes :—

" In one respect Mr. Bolton King broke entirely fresh ground. His scheme was the first in which the men had any share in the control. He began indeed by simply promising them a share in the profits if any were made ; but at Michaelmas, 1883, the larger plan was put into operation at Radbourne, and a year later the second farm started its career at Ufton. At Radbourne Mr. King was both landlord and capitalist. He let the land to the men at a rent of £1 an acre, and the principle of the scheme was that after this rent was satisfied, and six per cent.—subsequently reduced to five—paid on the capital advanced, any profit remaining should be divided, in proportion to wages, among the men working the farm and forming the association. It was arranged that the general business of the farm should be managed by a committee of three. Two of these were elected by the men, and the third was the manager, who was appointed by Mr. King. All orders must come from the manager alone, but in matters of importance he was expected to consult with the committee. All questions relating to the general position of the Association were to be decided by the com-mittee or the whole body of members. The Radbourne farm is 346 acres—heavy clay land, fertile but difficult to work, and suffering much from wet seasons. The grass land is good dairy land; the arable portion is suited for wheat and beans. It is not a good farm for sheep. The Ufton farm is larger—387 acres, of which a third is stiff clay and the rest light land—suitable for barley and turnips. It is less productive than Radbourne, but may be worked somewhat more cheaply, and suffers less from wet weather. The Ufton farm belongs to Balliol College, Oxford, and therefrom it was rented by Mr. Bolton King's arrangement. It has been managed on exactly the same principles as the Radbourne farm; but the rent is 17s. instead of a £1 per acre. One manager has acted for both farms. It should be added that each was in a bad condition when taken over.

"One feature of the experiment must not be forgotten. The wages paid to the labourers on the farms—about

twenty all told—were two or three shillings a week in excess
of the average wages of the district. This fact, which has
been a good deal criticised, may, of course, be justified on
moral or social grounds; but Mr. King is quite prepared to
defend it on economic principles. He is convinced that
the mere physical results which the higher wages produce
make them remunerative. Then, again, in harvest times all
the farmers pay higher wages; it is simply in the winter
months that the scale on the Association Farms becomes
better than among the neighbouring cultivators. Economy
has also been effected by putting the men on short time
when work was slack, and giving them, individually, allotments to cultivate in the leisure thus obtained.

"Thus far the constitution of the system on which the
farms were worked. As for their financial history, is it not
written too legibly in the balance sheets which have from
time to time been published? Mr. King produces them,
explaining that in 1886 and 1887, during his illness and
absence from home, the method in which the accounts were
kept made it impossible for the auditors to prepare a
balance sheet. These financial statements, with the exception of that for last year, have already been made public
and been well discussed, and only a word or two about
them is necessary now. At Radbourne, it will be seen,
the first year was the only year of profit. Then the gain
amounted to £57. In 1885 there was a loss on the year's
trading of £612. This the committee mainly accounted
for on the ground that there had been a fall of from 30 to
40 per cent. in the value of live stock. The 1888 balance
sheet showed a loss for 15 months of £206, and the last
financial statement, not before made public, shows that the
year 1889 resulted in a further loss of £211. At Ufton,
financially, matters have been as at Radbourne—only more
so. The first year's trading ended in a loss of £387; and

in 1888 there was a deficit of £193. The loss for last year was £297.

"Mr. King is asked how far the inexorable logic of these balance sheets has influenced his views on the general subject of Association Farming. 'I think,' Mr. King replies, 'that, so far as it is possible to draw a generalisation from two cases, that things in this country are not ripe for Association Farming. In the first place it is not easy to find many managers with all the necessary qualifications for the work. You want men who have sufficient agricultural knowledge, who are perfectly honest, who are in complete sympathy with the labourers, and themselves take an interest in the scheme. The second drawback that has impressed me is that you do not get the same amount of interest where the men do not put their own money into the concern. The prospect of having a certain amount of profit to divide at the end of the year does not appeal to them as they would be appealed to if the alternative was a chance of losing their savings. We shall have to await another generation before there will be any possibility of investments in farming being made to any extent from the labouring classes themselves. By that time the village co-operative stores, for instance, will have saved a considerable amount of capital, and the money may be invested in agriculture.' 'The third difficulty,' Mr. King continues, 'is that productive co-operation presupposes habits of organisation; and, probably, it must be preceded by a considerable period of Trades Unionism. Even, however, were Association Farming generally adopted, I think very likely it would be brought about in a rather different way from the one we have been trying. It will come in a simpler and more gradual fashion, through co-operation in allotments, for example. The allotment holders will see that it is to their interest to combine for certain specific purposes—for buying

manures, hiring machinery, and in self-defence. The allotment committees already established in some of the villages
are educating the people up to this; and in the end the
co-operative principle in agriculture will come quite naturally
and almost unseen.'

"'And now to summarise shortly the direct causes of
failure.'

"'I must first say,' observes Mr. King, 'that the first
four years during which the farms were tried were not a fair
sample of what association farming may do. This was
owing to certain factors in the situation which made success
practically impossible. I must also say that failure in my
opinion cannot be put down to bad seasons or low prices.
But as to the direct causes of our want of success, I think
they have been something as follows : (1) The farms were
in a more or less bad state in the years 1886-1887. The
stock deteriorated, and the interests of the association
generally suffered. (2) The charges incident to co-operative
farming are very heavy. There are five per cent. interest
on capital, and two per cent. wages of superintendence,
neither of which would necessarily fall on the ordinary
farmer, working on his own capital, and which form together
a heavy leeway of seven per cent. for extra profits to make
up. (3) As I have said in summing up the general impres-
sions which these experiments have left on my mind, the
labourers having no capital of their own invested do not
show the same keenness of interest that they would if their
own savings were at stake, and it was not only a question
of missing a prospective profit, but they actually stood to
lose. The work on the farms has been careful and good ;
but it has not shown that superior zeal and enterprise which
are necessary to make co-operative farming pay.'

"In reply to a further question or two Mr. Bolton King
says that the Association's tenancy of the Radbourne farm

will terminate at Michaelmas, and their occupation of the Ufton farm as soon as the necessary arrangements can be made with the landlords.   As the interview concludes Mr. King reminds our representative that last year there was a small profit on Lord Spencer's Association farm in Northamptonshire ; and the general impression left on one's mind is that though the Radbourne and Ufton farms have not succeeded, the question—Is Association Farming a Failure? —should not be hastily answered in the affirmative; that it deserves some further trial before it is cast into the ' waste paper basket of universal philanthropy '; and that, however widely the methods will need change, the associative principle in agriculture is in itself a valuable one, and has a future before it."

The financial statements for the Radbourne and Ufton farms for last year, as audited by Messrs. Fisher and Randle, of Birmingham, now first published, are appended :—

RADBOURNE MANOR FARM ASSOCIATION.

Trade Account—December 31st, 1888, to December 31st, 1889.
Dr.

		£	s.	d.
To Stocks, etc., at December 31st, 1888, as per valuation of Mr. S. P. Graves		2,726	14	2
To Purchases, viz.—				
Sheep	41 18 0			
Cattle	189 15 0			
Pigs	43 8 6			
Seed	24 12 1			
Corn, cake, and manure . . .	13 13 3			
		313	6	10
Rent, rates, taxes, and tithes		393	16	0
Labour (including manager)		552	10	0
Steam ploughing, tradesmen's bills, and sundries . .		194	11	3
Bank charges		1	13	10
Interest		140	0	0
Total		4,322	12	1

Cr.

		£	s.	d.
By Stocks, etc., at December 31st, 1889, as per valuation of Mr. S. P. Graves.		2,590	7	11

Sales, viz.:—

	£	s.	d.
Sheep	35	6	6
Cattle	566	5	6
Pigs .	113	17	4
Wheat	223	3	9
Corn	212	12	0
Straw and Hay	45	5	6
Dairy and Poultry	283	19	2
Sundries	40	3	3

	£	s.	d.
	1,520	13	0
Loss for twelve months .	211	11	2
Total .	4,322	12	1

The balance sheet shows:—Liabilities: To cash creditors, holding mortage debentures, £2,835 ; sundry creditors (unsecured), £259 1s. 11d.—£3,094 1s. 11d. Assets: By stock as per valuation of Mr. S. P. Graves, £2,590 7s. 11d.; sundry debtors, £53 8s. 4d.; cash at bank, £32 10s. 2d.; loss from October 1st, 1887, to December 31st, 1888, £206 4s. 4d.; loss for year ending December 31st, 1889, £211 11s. 2d.—£417 15s. 6d.;—£3,094 1s. 11d.

UFTON HILL FARM ASSOCIATION.

Trade Account, December 31st, 1888, to December 31st, 1889.

Dr.

		£	s.	d.
To Stocks at December 31st, 1888, as per valuation of Mr. S. P. Graves		2,540	2	1

Purchases, viz.—

	£	s.	d.
Sheep	43	15	0
Cattle	8	10	0
Pigs.	10	0	0
Seed	29	17	10
Corn, cake, and manure .	98	15	6

	£	s.	d.
	190	18	4
Rent, rates, taxes, and tithes .	422	17	2
Labour (including manager) .	507	3	1
Tradesmen's bills and sundries	96	9	8
Interest .	129	3	10
Total .	3,886	14	2

Cr.

	£	s.	d.
By Stocks, December 31st, 1889, as per valuation of Mr. S. P. Graves	2,235	11	2

Sales, viz.—

	£	s.	d.			
Cattle	346	18	6			
Sheep	252	3	11			
Horses	29	0	0			
Pigs	128	11	0			
Wheat	292	19	9			
Corn and Barley	93	1	1			
Straw and Hay	120	18	9			
Dairy, poultry, and eggs . . .	56	19	7			
Sundries	32	17	8			
				1,353	10	3
Loss for year				297	12	9
Total				3,886	14	2

In this case the balance sheet shows :—Liabilities : To cash creditors holding mortgage debentures, £2,650; rent, £97 5s.; sundry trade creditors (unsecured). £58 17s. 10d.; £2,806 2s. 10d. Assets : By stocks, &c., as per valuation of Mr. S. P. Graves, £2,235 11s. 2d.; sundry debtors, £58 3s.; cash at bank, £20 18s. 4d.; loss for year ending December 31st, 1888, £193 17s. 7d.; loss for year ending December 31st, 1889, £297 12s. 9d.—£491 10s. 4d. : £2,806 2s. 10d.

CHAPTER XI.

A COLLIERS' COW-CLUB.

"Through smoke clouds rising thick and dun
    As dust of battle o'er us,
Their white horns glisten in the sun
    Like plumes and crests before us.
In our good drove, so sleek and fair,
    No bones of leanness rattle;
No tottering hide-bound ghosts are there,
    Or Pharaoh's evil cattle.
Each stately beeve bespeaks the hand
    That fed him unrepining;
The fatness of a goodly land
    In each dun hide is shining."
                                        WHITTIER.

GREEN grass and some knowledge of stock
would seem to be the two indispensable pre-
requisites for successful cow-keeping, and
neither one nor the other would one have
expected from the nature of things to abound
in the neighbourhood of Newcastle. Certainly
it does seem surprising that we should have to
go to the heart of the English coalfields to find
a successful instance of a Co-operative Dairy
Farm. And yet in the village of North Seaton

there is a little band of resolute north country colliers who have found time, in addition to their regular work in the mines, not only to undertake the management of some thirty cows, but actually to pay an annual dividend of over 7 per cent. as a result of their enterprise in dairy farming.

The North Seaton Co-operative Farming Society.

The North Seaton Co-operative Farming Society is in reality a Co-operative Cow Club, established ten years ago, in the very heart of the colliery district, by Northumberland pitmen.

I am indebted for the following interesting account of the origin of the Society to my friend Mr. J. Pringle, a member of the Club, and until lately a working pitman in that district :—

A Cow Club " down among the Coals."—" Eleven years ago the colliery village of North Seaton, Northumberland, had a severe visitation of fever. Sad havoc was played among the little ones of the village. Almost all the children were more or less affected by this pestilent visitor. The doctors emphatically impressed on the minds of parents to give their children a better supply of milk—as much as they could drink—and to have it pure. To attain this there were many difficulties. · Two chief ones—milk was scarce, and its scarcity didn't warrant its quality. The workmen of the village, as sound practical men, saw their position, and their good common sense told them that if they could get cows of their own they could have as much pure milk as they needed. A meeting of the workpeople of the village was called, and the question discussed, and it was finally agreed to 'commence at once' with a Co-operative Dairy Company. It really began life in August 1872. Fifty-four people signified their intention to become members, and a capital of £84 10s. was raised on shares of £1 each. The employers promised to let the co-operators have the land at the same price they themselves were paying for it, and also promised to erect all the buildings at a mere nominal cost.

"Seven full-bred cows were got to commence with, but one great difficulty at this early commencement was the question of water. The men themselves sank a deep well, but owing to the nature of the strata through which it was sunk the water would not stay in the well, but leaked out. It is even now a source of much trouble, so much so that very often the owners of the collieries have to lend water for the society, which is always done free. Hay, manure, etc., are all conveyed to and fro by the employers' horses and carts free of cost, also the dairy man and woman are by the Coal Company allowed house and coals free. At the end of the first half-year a dividend of 10 per cent. per annum was declared, when a call of 5s. per share was made, and the number of members increased by 63, which made a total of 117. The capital now increased to £220 6s. 6d. For three years the society each year paid a dividend of 10 per cent., while their number of cows had increased to 13.

" It was here imagined by the committee that 'half-breds ' would feed on the land at disposal better than ' full-breds.' Accordingly the committee resolved to dispense with the latter and purchase the former, which resolve was carried into effect. ' I think,' says my informant, ' this was a grave mistake, as all our " half-breds " are being disposed of and we are as fast as possible getting back "full-breds."'

"Since 1875 the dividend has averaged 7 per cent. per annum. But even this dividend has been got under the most trying difficulties—difficulties strong enough to break down any ordinary enterprise of the kind. Each year the society sustained the loss of a cow, and from '77 to '79 the colliery was entirely ' laid in.' This caused large numbers of workmen to leave the village, which diminished the demand for milk. However, nothing daunted, the remaining members set to work and sought out new markets for their milk. A market was got at a town four or five miles away, to which town the society sent forty quarts daily. This energetic step enabled the co-operators to tide over the two years of the colliery's standing still with success. When work recommenced in the village, the inhabitants again took all the milk. To both members and non-members milk is sold at threepence per quart, pure milk. The society has not a monopoly, as other milk vendors come to the village.

" And now, amidst all the fearful depression of trade, this noble little enterprise has held up its head and flourished, notwithstanding the promoters of the Dairy Society were pitmen. The great cause of success has been earnestness and purity of purpose. If these qualities had not prevailed to a great extent, you may depend that instead of now numbering 108 members and owning £222 18s., and 13 cows and other property, the North Seaton Dairy Society would have been among the

> "' Little systems that have had their day,
> Had their day and ceased to be.'"

The following is a copy of the nineteenth half-yearly balance sheet issued last July :—

CASH ACCOUNT.

Receipts—June 30th, 1882.

		£	s.	d.
To Cash in hand		6	2	$3\frac{1}{2}$
Milk money		170	5	$5\frac{1}{2}$
Cows sold . . £77 2 0				
Calves sold . . 6 0 0				
		83	2	0
		£259	9	9

Payments.

		£	s.	d.
By paid for cows		83	13	0
Provender		71	3	8
Manager's wages . £32 19 0				
Other wages . . 2 4 0				
		35	3	0
Printing and stationery . . .		0	12	0
Railway carriage		3	6	0
Incidentals		4	4	1
Withdrawals . , . . .		2	5	2
Manure for fields		8	10	$7\frac{1}{2}$
Committee's expenses . . .		1	6	6
President		0	10	0
Treasurer and secretary . . .		3	10	0
Auditors		0	10	0
Rent of room		0	5	0
Balance in hand		44	10	$8\frac{1}{2}$
		£259	9	9

CAPITAL ACCOUNT.

Liabilities.

	£	s.	d.
To members' claims . . .	222	15	3
Reserve fund	26	17	11½
Rent of fields	33	10	0
Balance of profits	17	0	0
	£300	3	2½

Assets.

	£	s.	d.
By value of cows	192	18	0
Goods in stock	45	9	9½
Fixed stock	12	8	0½
Outstanding accounts . . .	4	16	8½
Balance in hand	44	10	8
	£300	3	2½

PROFIT ACCOUNT.

	£	s.	d.
To interest on £220 15s. 3d., at 5 per cent. per annum . . .	5	11	4¼
Reserve fund	7	0	0
Reduction of fixed stock . . .	4	8	7½
	£17	0	0
	£	s.	d.
By balance of profits . . .	17	0	0
	£17	0	0

When one reads an account like this, it certainly does make one hope that the time is not far distant when the labourers and cowmen.

of the finest dairy districts in England may learn to follow the example of the Northumbrian colliers.

When one remembers, moreover, that we have high medical authority for the statement, that notwithstanding the higher wages and in other respects the undoubtedly improved position of the rural labourer, his bodily physique has steadily of late years deteriorated, owing very greatly to the loss of milk as a regular article of diet, one can well understand with what manifold advantages, not merely economical, any system would be fraught which would revive once more the custom of cow-keeping, at one time almost universal amongst the English peasantry. It is altogether most lamentable that at the present time large numbers of the children of farm labourers should be brought up scarcely ever tasting the natural diet of childhood.

One is glad, however, to know that there are parts of England where this subject is receiving from the landlords that attention which is its due, and where efforts have been successfully made to introduce cow-keeping by the labourers. I am indebted to Lord Tollemache, of Helmingham, for a most interesting pamphlet on this subject, by Mr. Henry Evershed, reprinted from the Royal Agricultural Society's Journal for 1879. From that paper I gather that on the Helmingham Estate cow-keeping by labourers has been for several years eminently successful, and entirely free from the drawbacks which the opponents of peasant farming of any kind are only too ready to discover and to magnify into insuperable difficulties. This is Mr. Evershed's report :—

" The Helmingham Estate lies in the middle of the Cheshire dairy district. The farms average about 200 acres each, and about three men—cowman, horseman, and labourer —are employed on each. Nearly all the labourers, as well as some of the small tradesmen on the estate, keep cows. There are about 300 cottages ; at the time of my visit 260 of the cottagers were cow-keepers, and before the close of the year seven others were to be added to their number. Any man who finds himself in a position to keep a cow is enabled to do so by an allotment of pasturage to his cottage. About three acres suffice for the keeping of a cow, of which about one acre is mown, one-quarter of an acre is in tillage, and the rest is pasture. The rent of the land is the same as that of the adjacent farming land. Generally speaking, the three acres required for a cow are attached to the cottage, but in some cases a pasture is set apart for cottagers' cows, and 30*s.* per cow is charged for grazing. The organization of the system is perfect. The butter is collected and marketed by small dealers, residing generally on the estate and being themselves cow-keepers."

To this information may be added the testimony of Mr. Stephen Crawley, of Tarporley, Lord Tollemache's agent :—

"The labourer has, comparatively speaking, a plentiful board for his family, and at a cheap rate.   He has his cow, pig, and land to occupy his spare hours, which might otherwise be spent in the beerhouse.   His family have an opportunity from their infancy of taking part in the management of stock, and this is most necessary if they are to grow up into thorough stockmen.   The chief benefit derived by the farmers is that when they have the nomination of tenants of cow-keeping cottages, they can obtain the best and most intelligent men, who, but for the advantages of a cow, would drift into the large towns."

I regret that I am not able to give *in extenso* the Rules of the excellent Cow Insurance Clubs which form so important an element in the success of the system.   In his letter to myself on this point, Lord Tollemache says—" If you do establish this system, which answers admirably, you ought to urge those who keep cows to establish cow clubs.   The system of begging for assistance in case of the death of a cow is very objectionable indeed.   As a proof of how well these clubs answer, worked entirely by the labourers themselves, I may tell you that I never by any chance have an application for assistance owing to the death of a cow."

Similar evidence to that given above concerning the Helmingham Estate has been collected by Mr. Evershed from other parts of the country. "Nowhere," he says, "have I met with a single objector to the system of cow-keeping by labourers among any persons who are practically acquainted with the system, though some have objected to it who have never seen it."

CHAPTER XII.

A CO-OPERATIVE COW CLUB IN BUCKS

"Hi! diddle, diddle, the cat and the fiddle,
The cow jumped over the moon."

ABOUT sixteen months ago there met in my library at Granborough some score or so of agricultural labourers, who had come together to see what could be done towards fairly launching themselves as a society upon, to them, the unknown seas of co-operative enterprise. I need hardly say to those who know anything of the cautiousness, not to say wary stolidity of the Bucks rural mind, that it had not been without a good deal of effort, spread over a long period of time, that this point had been reached. Ten years ago, when I first began work in the county, I did not then appreciate, as I trust I do now, the fact that the stability of an institution is very much in proportion to the slowness of its growth to maturity, and so, in the inexperience

of those days, did not see that the step from
the conception of a good idea to its realisation
was not likely to be so short as I had imagined.
And such a good idea I thought I had in co-
operation, as a principle to be applied, by
associated labourers, to the trade of cow-keeping
and small dairy-farming.

Co-operative Dairy-Farming.

In the year 1872, in a lecture on "The Labour
Question" (then becoming a very burning
question indeed, owing to the agitation of the
agricultural labourers, so nobly generalled by
Joseph Arch), and which was afterwards republished in my "Village Politics," I ventured to
make the following suggestion :—" Some application of the principle of co-operation, if
political economists tell us rightly, is sure to be
tried in the future in agriculture, as elsewhere.
Why, then, I would ask, should not the experiment be made in the direction of associations
among the labourers, not only for the purposes
of strike, or a forced rise in wages, but for purposes of direct production ? Would it not be
possible to establish in the agricultural districts,
such of them, at any rate, as are chiefly pastoral,

societies organised upon co-operative, or, more strictly speaking, joint-stock principles, by which cow-keeping, dairy-farming, in fact, on a small scale, might be successfully undertaken by the associated labourers ? Such an experiment, if carried out with success, would certainly be most valuable, as tending to develop habits of self-reliance and self-government in village communities. It is by experiment only that the final and satisfactory solution of any problem is ultimately reached. And for this reason, if for no other, it appears to me the experiment I have suggested is, at least, worthy of trial."

Difficulties of Propagandism.

The same idea, with more detail, I continually pressed upon my labouring parishioners, in season and out of season, until I fear, to some of them, " the Parson with Co-op. Cows on the brain " became rather a bore than otherwise. However, parsons, I suppose, get used to being thought bores, and so I went on preaching. My most opportune occasion for making co-operative converts was always in the winter months, when I was accustomed to meet the men for night-school work and talk about things in general, and especially at the annual supper of the tenants on my glebe allotments. Here, by virtue of telling them such inspiring stories as that of the Rochdale Pioneers, and how " the famous Twenty-eight," beginning with twopence a week subscriptions, commenced business in 1844 with twenty-eight members and £28 capital, had increased the following year to seventy-four members, with £181 capital, having made £22 profit on business done to the extent of £710 ; and in 1876 to 8,892 members, with £254,000 capital, having made £50,668 profit on business done to the extent of £305,190, I occasionally succeeded in making a few converts, who, however, generally " lapsed " (at any rate from any desire for practical experiment) when the spring field work put an end to winter talks.

Many a time I feared that all my preaching was going to be in vain. I had not then, as I said, learnt the lesson of the parable either of the oak-tree or Jonah's gourd.

"*Spell it wi'h an S.*"

In the early part of last year, however, I

was lucky enough to persuade Miss Hart, who has been doing such good work lately for co-operation and the principle of participation in profits among the workmen's clubs of London and elsewhere, by her lectures, on the French house-painter Leclaire, and his scheme of co-operative industry, and on the Irish agricultural co-operative estate at Ralahine, to come down to Granborough to deliver to the labourers this latter lecture. She had a crowded room and an enthusiastic audience, and the pathos and generous sentiment with which she told that story of noble effort and self-sacrificing endeavour, with, alas! its too sad ending, fired their hearts with a desire to make trial of the new principle. A committee of some half-dozen or so was chosen to draw up rules for the proposed association, and after some unavoidable delay in procuring the necessary Government Registration, the Society was fairly ready to be launched. At the meeting to which I have already alluded, it was found that there were some twenty-three members ready to join as 10s. shareholders, but who were not in a position to pay up at once the whole amount. Some eight or nine pounds for the present was, in fact, all that we could command by way of capital. What was to be done? That would not buy even one co-op. cow. A brilliant thought, however, struck one of the men, worthy of the immortal "Spell it with a wee, Sam ny, spell it with a wee!" of the elder Weller. "Spell it with an S, sir," said one of the men—"spell it with an S!" And so we did. For Co-op. "Cows" we wrote Co-op. "Sows." On the 7th June the Granborough Co-operative Association fairly opened business as the proprietor of two fine brood sow pigs; and, to cut a long story short, here was I in October its proud President, just returned home from my holiday, greeted almost as I entered the village with the stirring news, "Please, sir, the co-op. sows have got eighteen little pigs!"

The Poetry of Co-operative Pig-keeping.

Talk of co-operation and enthusiasm! here's co-operative enthusiasm indeed! A fortnight ago the twenty-nine members of the society could only be set down as the owners of scarcely the fifteenth part of a pig between them, and now each member may claim almost a pig apiece. O wonderful dispensation of nature!

O beneficent principle of increase! O glorious virtues of co-operation! Talk of poetry. Why, I believe each of us at this crisis in our fortunes felt himself a very Herrick, ready to sing as he did of the

> " Plenty dropping hand
>   That soils my land,
> And giv'st me for my bushel sown
>   Twice ten for one ;
> Thou makest my teeming hen to lay
>   Her egg each day ;
> Besides my healthful pigs to bear
>   Thrice nine each year,
> The while the conduits of my kine
>   Run cream for wine."

The Prose of the Matter : Rules and Regulations.

But to return to the prose of the matter. The rules of the society, which is registered under the Industrial and Provident Societies Act, are divided into two sets—special and general. The general rules, numbered 1 to 144, are those published by the Central Board of the Co-operative Union, and approved by the Government Registrar. I should strongly re-commend any proposing founders of village co-operative societies to adopt these " Model

Rules" in their entirety. They may easily do this, whatever the special objects of their society may be, by means of special rules, which the general secretary, Mr. E. V. Neale, will prepare for them, if applied to at No. 2, City Buildings, Corporation Street, Manchester. The genera rules are kept in stock, and sold at 2*d.* a copy, for a book of fifty pages, so that a society may get a hundred copies made up with its own special rules for little more than £1. The special rules of the Granborough Co-operative Association, Limited, were drawn up by myself, in consultation with our committee of seven labouring men, and were kindly revised by Mr. Neale. They are as follows :

Granborough Co-operative Society.—"I. General Rule 3. —The name of this society is the Granborough Co-operative Association, Limited.

" II. General Rule 3.—The objects of the society are to carry on the trades of cowkeepers and dairy farmers, and of general dealers both wholesale and retail.

" III. General Rule 4.—The registered office of the society shall be at the Parish Room, in the Vicarage of Granborough, in the county of Buckingham.

"IV. General Rule 22.—The shares shall be of the nominal value of ten shillings, of which four shillings shall be paid on allotment, and the remainder by instalments of not less than threepence per week.

"V. General Rule 27.—No member shall hold more than one hundred shares.

"VI. General Rules 34 and 35.—Repayment of shares· These rules shall apply to all the transferable shares of the society.

"VII. General Rule 37.—No withdrawable shares shall be allotted except on the resolution of a majority of two-thirds of the members present at a special general meeting.

"VIII. General Rule 68.—The Annual Meetings.—After the first general meeting there shall be only one ordinary business meeting in each year, which shall be held on the first Monday in May.

"IX. General Rules 85, 86, and 101. The Committee shall consist of the president, vice-president, treasurer, secretary, and five committee-men, of which nine members three shall retire at each annual meeting.

"X. General Rule 93.—The Committee shall meet on the first Monday evening in each month.

"XI. General Rule 110.—The seal of the society shall have the device of a milking stool, with the motto *molliter mulce.*

"XII. General Rules 127 and 141.—Application of Profits.—The net profits of the society, after providing for rent, rates, taxes, interest and redemption of loans, insurance, and all expenses of management, shall be apportioned by the annual meetings in the following manner :—

(*a*) In forming a depreciation fund of ten per cent. on all property of the society.

(*b*) In paying such dividend on all fully paid-up shares as may be determined by each annual meeting.

(*c*) In forming a " Public Good " Fund, to promote the moral and intellectual welfare of the members, and to disseminate a knowledge of the principles of the co-operative faith.

Of the General Rules the following are those which relate to the conditions of membership:—

" 9. Payment on Application for Membership.—Every person shall pay, on applying for admission, the sum of 1*s.*, for which he shall receive a copy of the rules.

" 11. Application by an Individual.—An application by an individual may be made either in the form contained in Rule 140 (given below), or by making such payment as is required upon an application for admission, and the signature of a receipt for a copy of the rules in the form contained in the said rule, either by the applicant or on his behalf.

" 12. How Applications shall be dealt with.—Every application for admission shall be considered by the committee at its first meeting after it is made, or so soon thereafter as is practicable ; and if it is approved, the name of the applicant shall be entered on the list of members and the register of shares hereinafter mentioned, for the number and description of shares required to be held by the rules, or any larger number applied for and allowed to be held thereby, upon such approval and such confirmation thereof, if any, as the society may direct.

" 13. Notice of Refusal or Admission.—Notice of the refusal of an application, or of the entry of the name of any applicant on the list of members, signed by the secretary, shall be sent to the applicant, at the address mentioned on the form of application or receipt, within one week after such refusal or entry is made.

" 14. Repayment of the Entrance Fee.—The sum paid on an application for admission shall be repaid on demand to the person by whom or on whose behalf it was or purports to have been paid, within one calendar month after the date of the said notice in the cases following :—

(1) If the application is refused ;

(2) If it is not granted within three calendar months after the application ;

" (3) If it is based on a payment made on behalf of any person without his authority.

" 140. (1) Application for Admission by an Individual :

" (*a*) By Application for Shares—I, the undersigned, hereby apply for　　[*transferable*] shares in the above-named society, in respect of which I agree to make the payments required by the rules of the society, and otherwise to be bound thereby.

" Signature of applicant, stating his address and occupa·tion."

"(*b*) By Payment for a Copy of Rules.—Received this day a copy of the rules of the above-named society, for which I have paid [*on account of the undermentioned applicant*] the sum required to be paid on an application for admission to the society.

" Signature as above, or if the payment is made by any other person than the applicant, of this person, stating the name, address, and occupation of the applicant."

Members of our society naturally had to put

up with a good deal of mild chaff from the wits of the neighbourhood, which I am bound to say they have hitherto borne with a very commendable degree of equanimity. Here is a specimen, addressed to the President about Christmas time.

Hi! diddle, diddle!

"Well, Vicar, how does the cow-club get on? You have not turned the 'S' into a 'C' yet, I hear. But no doubt the club prefers bacon rashers to roast beef for its Christmas dinner. By-the-bye, as the Co-op. Cow still seems 'up in the air,' I wonder you don't change that very humble device of the milking stool on your corporate seal. What do you say, now, to the 'Cow that jumped over the moon,' with the legend 'Hi! diddle, diddle!' for a motto? That would surely be more ambitious and appropriate. Well, good morning! a merry Christmas to the Co-op. piggies!"

Those may laugh who win. We did not forget, luckily, that there has always been subject for ridicule in the early chapters of the history of all new social experiments, and that

the co-operative movement has been no exception to the rule. When Robert Owen
opened the first Labour Exchange in London it
was an easy matter to laugh at "the greatest
philanthropist in the world, the correspondent of
all the monarchs of Europe," as a retail dealer
in pink-eyed potatoes, short dips, and treacle.
There was subject also for raillery, no doubt, in
the barrowful of provisions with which the Rochdale Pioneers first set up shop. All the
"doffers" of Toad-lane were in a roar when this
humble stock-in-trade was first exposed to view.
But the Pioneers have the laugh on their side
now, when "the owd weaver's shop" has become
a large Central Store, "a commanding pile of
buildings which it takes an hour to walk
through," situated on the finest site in Rochdale,
overlooking the Town Hall and Parish Church,
with a turnover of £270,000 a year, and an
annual net profit of more than £30,000.

And so we Granborough co-operators took
heart to face even worse troubles than nursery
rhyme banter. Not that we have any cause as
yet to lose heart. If our progress has not been
brilliant it has been steady. When we commenced business a little over sixteen months ago

we had 23 members and a capital of £8. In 1882 we had got to 44 members and a capital of over £30. At our committee meeting on November 29th we found upon balancing the accounts for the half-year that we had made a net profit of £5 7s. 11d., on a total expenditure of £18 14s., or more than 25 per cent. on the capital invested. These results we are disposed to think were not unsatisfactory, and although we are all still anxious to see Co-operative Cows grazing in the Granborough meadows, our present experience of pig-keeping you may be sure is at any rate not likely to lessen our respect for the co-operative virtues of that "gintleman what pays the rent," as the Irish say.*

An "ex-Irish Landlord" writes to the *Times:*—"Mr. Bright says, 'It is obvious to any one who knows anything of these matters that capital employed in agriculture in any country must be absolutely unprofitable to the cultivator if he has to pay a rate of interest such as 24 per cent.' For Mr. Bright's benefit I hope you will let the following story appear, for the truth of which I can vouch. A clever little Irishman, then a tenant of my own, told me that 2s. 6d. per month was often charged as interest upon £1 sterling, and said that he had himself borrowed £1 on those terms—viz., 150 per cent. per annum. 'How on earth, Johnny,' said I, 'did you do that?' 'Well, your honour,' was his answer,

Every one knows La Fontaine's story of little Perette going to market to buy eggs. The eggs are hatched into chickens. The chickens produce a pig. In time the pig becomes a calf, and the calf grows into a cow. It is this dream of Perette that we Granborough co-operators hope to realize.*

Whether we shall succeed or not I cannot say. But if we do not, then I trust some one else will. For of this, at least, I am quite sure, that even failure will not disprove the principle of co-operation.

Conclusion.

Indeed, I venture to say, finally, that if we are ever to succeed in raising the platform of Industrial Morality, and healing what is at present, I fear, the standing feud between Capital and Labour, it must be by the substitution of the principles of co-operative

---

'when I got the pound I bought a bhonnal (a little pig) for 15s. ; with the other 5s. I paid two months' interest, and before the third month was out I sold my pig for £3, and paid off the debt and interest, and had £1 17s. 6d. in my pocket.'"

* "*June* 10*th*, 1883. Two heifers bought, £21 10s.' —*Extract from Books of the Society.*

faith for that spirit of competitive selfishness which now forms the orthodox foundation o economic science. The ideal of co-operation is, indeed, a noble one, for it means "the transformation of human life, from a conflict of classes struggling for opposite interests, to a friendly rivalry in pursuit of a good common to all : the elevation of the dignity of labour, a new sense of security and independence in the labouring class, and the conversion of each human being's daily occupation into a school of the social sympathies and the practical intelligence.'

APPENDIX I.

EXTRACT FROM THE *EAST ANGLIAN DAILY TIMES*, TUESDAY, DECEMBER 30TH, 1879.

"GARDEN FARMING IN EASTERN COUNTIES.

"The opening of a new vegetable, root, and general produce market at Stratford, in a densely populated part of London, and yet within easy reach from every railway in the Eastern Counties, renders garden farming a far more important local industry than it has hitherto been considered. Times are deeply depressed in the agricultural world, and great interest therefore attaches to an enterprize which affords a new opening for the capital and skill of the farmer who does not look down upon market gardening simply as the employment of the cottager. Hitherto the Eastern Counties, though favourably situated, have done little in this branch of cultivation, and vegetables sent to London have been chiefly consigned to Spitalfields and Covent Garden. The consequence has been, that the immense population on the northern and eastern outskirts of London have been as unable to obtain vegetables fresh, or in any quantity, as if there were no supplies sent up at all. The population of Stratford alone, where the new market is situated, is 120,000, and adjoining it are Bow, Bromley, Stepney, Limehouse, Canning Town, West Ham, Woodford, Forest Gate. Leytonstone, etc., containing in all a population of fully 500,000. When the immense amount of vegetables required for this market alone is considered, and the consequent almost unlimited demand borne in mind, it will easily be understood that the scope for market garden farming is very large. .The Great Eastern Railway Company have erected the Stratford Market on land adjoining their system, and were induced to take this course in consequence of the belief that, if the means of distribution were more perfect, the cultivation of garden produce would become more general. They have offered this branch of agriculture further encouragement, by greatly reducing the rates of carriage of garden produce to the Stratford Market from all places on their line ; and so well has it sold, both there and in the old markets, that ten per cent. profit on the outlay for cultivation is found to be a fair average.

" The Company look mainly for their profits to the large amount of traffic which it is anticipated the market will bring on to the rails, and have let their warehouses at a merely nominal rent, while at the same time no charges are payable by those who merely sell *ex* truck direct into cart. The vegetables forwarded to the warehousemen for sale have been put on the trucks at places in Lincolnshire, Cambridgeshire, Norfolk, Hertfordshire, and Essex, and up to the present time, have reached a total of 200 tons per week, in addition to which the salesmen have consigned direct to customers a considerable quantity of produce. A ready demand has been experienced for all that has been sent to the market, and it is hoped that when farmers and growers in the Eastern Counties become alive to the advantages of this branch of agriculture, and are aware of the facilities offered for the disposal of vegetables of all kinds, a greater supply will be sent up, as there is ample accommodation and demand for it.

"At the present time, large quantities of the commoner vegetables, such as potatoes, turnips, onions, cabbages, etc., are grown in Lincolnshire, Yorkshire, and so far north as Scotland, and sent up to the London markets, where, notwithstanding the cost of cartage to the station, railway carriage, and dealers' commission, they are sold at profitable prices. It is the fact that, given a piece of land not too far from a railway station, properly managed, and within 90 miles of London, or other large centre of population, it can be made to pay handsomely under a system of garden cultivation, and the farmers of the Eastern Counties have in this branch of agriculture one means of avoiding that stagnation and depression which exists now because of the failure of the crop which has been chiefly relied on for profit.

"In considering the subject, it may be well to point out a few of the leading items of expenditure, and the position which the various products hold in the London market. Potatoes, although a precarious crop by reason of their liability to disease and sensitiveness to the extremes of the weather, are yet found to make enough profit to encourage their growth in all parts of the country. Failure has, in many cases, induced farmers to abandon them as field crops, but inasmuch as they are one of the staple wants of the people everywhere, they cannot long remain in disfavour. A liberal application of manure under any circumstances is necessary, if a full crop is to be obtained. Farmyard dung, guano, and nitrate of soda are chiefly used, and although the 'home supply' of the former is necessarily limited, what is lacking can be made up by stable and cow-shed manure and garbage from the slaughter-houses of London, which can be obtained at about 8s. a ton, to which of course has to be added the cost of railway carriage.

It may be taken as a basis that manure will cost £15 an acre, and to this must be added that of seed and labour. Notwithstanding the quantities of potatoes that are brought from abroad, fair samples make from £5 to £8 a ton, and eight tons per acre is a good crop. In the winter the price advances to £10. White turnips will make 30*s*. to 35*s*. per ton in the London market, while carrots, which are deservedly esteemed and sought after, make 30*s*. to 50*s*. a ton. Although the latter roots are universally grown in gardens, they have not as yet attained to general cultivation as a field crop. There are certain practical difficulties connected with their culture on a large scale, but the precariousness in the growth of potatoes, turnips, and clover, and the consequent necessity for a greater variety of green crops both for men and cattle, entitle the carrot to increased attention as a field crop. It is on sandy and light loamy soils, and those of a peaty nature, that it is grown most successfully, and under these circumstances 15 tons per acre is an average crop, although with liberal manuring and skilful cultivation double the weight is sometimes obtained. Parsnips closely resemble carrots in culture and uses, but they possess advantages over the latter in being successfully grown on a much wider range of soils, and, unlike the carrot, it rather prefers those in which clay predominates. It appears to have received very little attention from cultivators, for it is not sent to the markets in quantities at all reaching the demand for it. There is always a demand for cabbages, and they are at present more generally cultivated than anything else. They sell at 5*s*. 6*d*. to 6*s*. 6*d*. per tally of five dozen, equal to about £60 per acre. Onions take a great deal of care in their cultivation, but when landed in the market will repay the outlay and toil necessary for their successful growth. The gross returns on Lisbon onions is between £60 and £70 per acre, while an average amount of over £150 per acre has been made in three years upon a large acreage of pickling onions in the neighbourhood of London. Besides these crops, there are peas and beans which can be successfully grown ; but there is much trouble in connection with picking them where labour is not abundant, and for this reason they are often a cause of annoyance. The aver ige crop of peas is about 150 bushels per acre, and the average price obtained in the market is 2s. per bushel, or £15 per acre. Coleworts are a good spring crop, and a gross return of £100 an acre may be made on them. Much profit may also be made out of the cultivation of celery, spinach, rhubarb, lettuces, beetroot, and red cabbages for pickling, mint, sage, parsley, and other herbs. The judges in the competition for the prizes offered by the Royal Society of Agriculture for market garden farms, state that by dint of putting this and that fact together, it was gathered that an average crop of radishes sown between the celery would make at least £28 per race, say 1,100 dozen bunches at 6d. per dozen ; and the celery planted in the rows between the beds of radishes would make from £50 to £60 per acre, taking an average of seasons, say from 1,000 to 1,200 dozen bunches at 12s. per dozen bunches. Again, take lettuces and celery together, an average crop of lettuces would be worth about £30 per acre. Or a crop of coleworts and cabbages grown with celery would be worth from £25 to £30 per acre, plus the value of the celery crop ; so that even if the expenses amounted to £40 per acre, there would still be a good margin of profit.

" Market garden farming may require a large quantity of labour, and be in consequence expensive to carry on, but the returns are correspondingly large, and it has the further recommendation of increasing the demand for its products as it becomes more widely extended, because it will put the power to buy into the hands of men both in towns and the country who have now barely the means of existence. This fact puts on one side the argument that if the practice of vegetable growing became more general, a supply would be created greater than the demands of the public warranted. It may be said that foreign competition will outdo the English growers as the American supplies have destroyed the profits on wheat cultivation ; but the perishable nature of vegetables gives the English growers a decided advantage, while the cost of production and the expense of freight will in all probability too heavily handicap the foreign producers, and leave the English cultivators a margin of profit. The carriage of vegetables which are imported from France amounts to £4 a ton, whilst the highest rate for an equal amount from any part of the Great Eastern system to the metropolis—that from Peterborough—is but 16s. 6d. Those sent from other countries, such as Holland, Belgium, and parts of Germany, have not to pay such high freightage as the French ones, but they do not arrive in such good condition in consequence of the greater length of the journey ; and the bulky and common vegetables, such as cabbages, turnips, and coleworts, are seldom sent from abroad, while by adopting some of the practices of foreigners for forcing—the bell-glasses used in the north of France, for instance—the markets could be well supplied early in the year. There does not seem to be any reason why men with moderate sized or small farms should not alternate ordinary crops with those of vegetables, and thus multiply the sources of their profit and increase their own and their country's welfare."

APPENDIX II.

THE FEARGUS O'CONNOR ALLOTMENTS AT MINSTER LOVELL, IN OXFORDSHIRE.

I am indebted for the following report upon the present condition of these allotments to my friend, Mr. T. Forster Rolfe, of All Souls' College, Oxford, who, at my request, kindly paid a visit to the estate last Christmas time. The causes of the failure of the original allottees under the O'Connor ballot are sufficiently evident from the first sentence of the Report: "The men that came in the first allotment were many of them ignorant of how to plant, it being reported of one that he asked what he was to plant to make bread; another sowed his turnips as thick as mustard and cress, and refused to thin them because they looked so flourishing; another wished to know how many *bushels* of the same seed to sow to the acre.

"*Its present condition.*—Out of the eighty holdings at present, not more than six are cultivated by their owners. The present position of the occupiers is therefore principally interesting from the point of view of the success of small holdings when rented by agricultural labourers or very small farmers. For it is not the case that every lot is held by a separate tenant or occupier—not a few lying next each ɔther, or even a little distance off, are rented by the same nan, and some of the few that self-own the one lot, rent other lots, and work them in connection with it.

"A man holding a four-acre lot has but little time to give to other work, if he gets all the good that he can out of his piece; still, on the other hand, such a holding will not by itself find sufficient to occupy his whole time. A three-acre lot, if anything, is worse than either of the other two, for this reason, whereas a two-acre lot is pronounced by the present occupiers to be far the most convenient holding of the three, as a man can take pretty regular work elsewhere, and, at the same time, find time to keep his own piece in a thoroughly profitable state of cultivation.

"*Quality of the soil.*—One of the most important considerations in the question of the Minster Lovell Estate, is the varying nature of the soil : some lots have stone brash, some clay, and some both together; if anything, the last is the most advantageous when rented, as the rental is not so high from the mixed character of the soil, whereas if the season is unfavourable to one part, it is probably favourable to the other. Those with a clay soil are certainly the least desirable.

"*Produce.*—The reason of this is, that the latter are the most uncertain soils in their returns to the potato crops, and this, as is natural in small holdings, is the most valuable article of production on the estate. Wheat and barley are the other chief productions, roots also being grown in small quantities for pigs (and here and there a cow), carrots occasionally; garden-stuff is grown just round the house; black crops are sown occasionally. I know only of one case now in which a cow is kept, though many attempted it at first. The man Jacobs who keeps it, in this case, owns or rents altogether 15 acres, and therefore has more opportunity of growing the proper stuff; with his four acres he originally tried it, but had to give it up."

" *Present Occupants.*—The present occupants are a mixed class,—one, a Mr. Radbon, is a baker with a pretty flourishing trade : he not merely bakes for himself, but also for such other holders as choose to send their dough made up to him, at the rate of $\frac{1}{2}d.$ per loaf. Others are pork butchers, some, as said above, by combining more than one lot, have their whole time occupied at home, but the larger number are those that work part of the time on their own lots, part of the time on those of the larger holders, or on one of the neighbouring large farms. It is worthy of note, that a large farmer (a Mr. Clack), with his farm adjoining, uses almost all his labour from off the estate and elsewhere. The occupiers generally have no difficulty in obtaining work from the fact of their being themselves small holders.

"The Union was at one time taken up by the present occupants, but now I think has no supporters there, but, on the contrary, there is rather a spirit of opposition to it.

" *Morality.*—Even within as short a space of time as ten years ago, the O'Connorville had a bad name, and was reported, though probably incorrectly, as not safe for passing through at night. That the original occupants were doubtful characters, and rather light-fingered, there is no doubt, for scarcely any man's property, a few years after its formation, was reported to be safe unless under lock and key ; but this class has dwindled more and more away, till within the last few years there is not one of them left. Nealy every householder now is reported to attend either Church or Chapel on the Sunday. The place is orderly at night, the occupiers are quiet people, and generally seem busily occupied with their land when at home.

" *House Accommodation, etc.*—The house accommodation

on the estate, as compared with that in modern labourers' cottages, is not good, though probably, as compared with those put up at the time of their erection, the cottages were equally good, if not superior. The cottage is of one story only, and oblong shaped, with three rooms—a sitting-room, and two bed-rooms, and accommodation for scullery, etc.

" *Buildings.*—They all have pigsties at the back, varying in number, and some a sort of barn, or building for putting straw, sacks of corn, or anything in ; some also have cart hovels, etc.

" *Influence of the Seasons.*—The present season had a greater effect upon the holders of the lots in Minster Lovell than upon small holders of allotments in other parishes. So much of the land being heavy, has influenced the potato, and so, indirectly, the number of pigs kept and killed ; many not even having the money this season to purchase pigs, as they otherwise would have done, to eat up their waste stuff. The most palmy days of the industrious holders were those when a railway was not yet opened at Witney, and they conveyed all their potatoes by cart to Cheltenham, a distance of about 26 miles, where they found a well-paying market for them ; many more pigs also were kept at that time than are now. The present, therefore, in one respect, is a bad time to judge of their general success, except in the case of those holding them in their own right. Still the case of a man named Norridge, originally a poor man from near Woodstock, who has made his money by his own lot, since purchased, and is now living on such money without further work, proves, to a certain extent, what may be done by them. Jacob, another man mentioned below, did not come to Minster Lovell till the age of 30, and at the age of about 55 owns 4 acres, and rents 11 more, with two horses, cow, and other farming stock. The following statistics, therefore, must be taken for what they are worth :—

"*John Jacobs.*—Four acres in own right, 11 rented at £31, 4 acres mixed land, 11 acres heavy clay, gave no statistics, except that he had lost, at present, on the three years that he had held the 11 acres.

"*History.*—At the age of 30, he took 2 acres in bad condition, he kept this on for one year, when he took 2 more also in bad condition, and so kept on for three years, when, in consequence of his rent being raised, he gave up the last two, and kept on the other two for ten more years. At last took four-acre lot on yearly tenure, at the end of first year bought it (for £350). This is his present holding, and is mixed land; gave up the two-acre lot, and put up buildings, etc., on his own; went on with his own lot for six years, making money on it. Three years ago took 11 acres of heavy land, on which he has lost since by the bad season. This last land he has drained, his landlord supplying tiles. He is also still, with the help of the same, putting up buildings, and improving place gradually. Notwithing the bad seasons, the land has been worked clean, though taken to in a bad condition; employs two boys of the age of 14 and 17, keeps two cows and two horses.

"Jacobs' history is a most remarkable one, as given by him in the *Labourers' Union Chronicle* (November 21st, 1874). He was brought up at Old Weston, Hunts, as a poor boy, in the hardest of circumstances. After much knocking about, and marrying (according to his account) to get a greater share of relief, he came to Minster Lovell, where, as shown above, he has gradually thriven ever since. He is now a thoroughly hard-working man, always turning his hand to something, His buildings are a picture of neatness and ingenious construction; the stones were all chopped by himself, and put together by himself on his own plan. The pigsties, especially, are perfect, and as solid as if built by the best mason; one of the disadvantages of the estates is a want of water for cattle, etc. Jacobs has remedied this by turning the quarry, from which he dug his stones, into a large reservoir for the drainage of the land, lining it inside with cement. Jacobs' cottage is the neatest that I have entered on the Estate : everything is perfectly clean and put away in its right place. His wife is like himself, neat and bustling, and always ready to see you at the cottage when she knows you. Jacobs is by far the most interesting and progressing of all on the Estate, and is owned by all, except those jealous of him, to be by far the best workman."

GRANBOROUGH GLEBE ALLOTMENTS.

Agreement *made this............day of....................
188 , between the Rev. CHARLES W. STUBBS, M.A.,
Clerk, Vicar of Granborough, on the one part, and
....Labourer, on the
other part, witnesseth that the said................................
hath agreed to take an allotment of land, No.................
containing half an acre of ground, for one year, from
Michaelmas 188 , at the yearly rent of £1 13s., subject to the
following conditions, which will be strictly enforced*

Conditions.

I. The Rent fixed by agreement is to be paid half-yearly
on the 26th June and the 26th December, to the above
named Rev. C. W. Stubbs, or such person as shall be appointed by him to receive it.

II. The tenant to quit on a half year's notice, terminating
at Michaelmas.

III. No allotment or part of an allotment to be sub-let.

IV. That upon the quittance of an allotment by any
tenant, the crop and good-will, if any, shall be subject to
valuation, to be agreed upon by two valuers, one appointed
by the outgoing, the other by the incoming tenant, and in
the event of their not being able to agree, by an arbitrator
appointed by the Vicar.

V. The allotments to be open to inspection at all times.

VI. The land is at all times to be cultivated with the
tenant's best skill and diligence, and in such manner with
respect to the quantity and quality of manure as shall be

satisfactory to the Vicar. Any appearance of neglect or bad cultivation will be considered sufficient reason on the part of the Vicar to give notice to quit.

VII. The holders of these allotments agree to prevent depredations on each other's land, and to assist in detecting and convicting persons who destroy or injure fences or crops of any description.

VIII. No footpaths are to be made across or by the side of any allotment. The roadway in each case is measured into the amount of each lot, and the tenant is responsible for keeping such roadway in repair, with the exception of the piece of roadway between the upper hedge and the spring, dividing the triangular lot, No. 20. This piece of roadway (measuring 4 poles) belongs to the Vicar.

IX. All fences and hedges belong to the Vicar. In no case is any hedge or tree to be cut without special consent.

X. The tenants of the Glebe Allotments have no right of way across Vicar's Leys. The tenant of that field has special instructions from the Vicar to keep all gates locked. In Harvest time special leave may generally be obtained from the tenant to take crops across that field, but it must be clearly understood that this is in no case to be done before such special leave has been granted.

XI. Should any tenant be convicted of theft or other misdemeanour, he will be liable to immediate ejectment, without notice : the crops to be left on a valuation, according to Rule IV.

XII. Any tenant infringing any of the above conditions to forfeit his allotment.

Signed this..............day of........................188 .

CHARLES W. STUBBS, *Vicar.*

........................*Labourer.*

APPENDIX IV.

RULES OF THE

BULKELEY COW CLUB,

HELD AT THE

PRIMITIVE METHODIST SCHOOL-ROOM,

BULKELEY, IN THE COUNTY OF CHESTER.

Name and Constitution, etc.

1.—This Society shall be called the "Bulkeley School Cow Club," and is established for the purpose of making the most effectual provision for casualities affecting its members in case of loss of cattle. It shall consist of a Treasurer, two Trustees, two Stewards, two Markers, a President, a Secretary, a committee, and an unlimited number of Members. From the date of these Rules, no *new* members shall be accepted, who may reside above four miles from the place where the Club is now held.

Time and Place of Business.

2.—The Society shall meet on the first Monday in every month, at the Primitive Methodist School-Room, Bulkeley: at eight o'clock, p.m., in the summer half-year, and seven o'clock, p.m., in the winter half-year.

Entrance and Contributions.

3.—Members who have joined this Club previous to the adoption of these Rules shall pay one shilling per head as entrance, and sixpence per head, per month, contribution. All new members joining after the date of these Rules shall

14

pay the sum of 1*s*. 6*d*. per head, entrance, and sixpence per head, per month, contribution, so as to have a full claim on the funds raised by its older members. Each member shall be supplied with a book of Rules at a charge of threepence. No member shall be allowed to enter more than six cows, and, should any member have more than that number in his or her possession, it shall be discretionary with the Marker which six he accepts.

Fines and Arrears, etc.

4.—Any member neglecting to pay the contribution for four nights, shall pay a fine of threepence for each and every beast entered with this Society, and, if he or she so neglect payment for five nights, to be excluded, unless otherwise agreed on by the committee.

5.—Each member shall pay up all arrears of contributions on the clearance club night in May, and shall also pay up their contribution on the first club night after the clearance.

6.—Any member having resigned or become excluded, and shall again wish to become a member, he or she shall be accepted if there is no former charge against them, on payment of the usual entrance fees, etc.

7.—No member shall be entitled to any claim on this Society, before the expiration of fourteen days after entering any cattle. Neither shall a member receive any benefit who is also a member of another Society of the same kind.

Appointment and Duties of Officers.

8.—The officers shall be elected annually, or old ones re-elected if eligible. The *Treasurer* shall keep the surplus stock of money from year to year, and shall pay all losses, etc., when called upon by members producing a cheque for amount of claim from the Secretary, and signed by a Steward or Stewards.

The *Trustees* of this Society shall invest in the National Provincial Bank such sum as may be considered necessary as a standing stock for the purpose of meeting any heavy losses it may sustain, such sum to be invested in the name of the Bulkeley Cow Club. The Trustees shall not withdraw any of the stock without the sanction of the committee but shall be empowered to receive the interest arising from stock yearly, and pay it over to the Club. The Trustees shall remain in office during the pleasure of the Society.

The *President* shall preside at all meetings of the Society, and, each club night, shall receive all money from the hands of the Stewards, and shall convey it to the Treasurer, together with a note from the Secretary certifying the amount.

The *Stewards* shall attend every club night, or in default be fined 1s. or such sum as may be decided upon, and shall receive from members all contributions, entrances, fines, etc.

The *Secretary* shall attend each club night, and enter all contributions, etc., in the pence book and on the members' cards. He shall also attend and take minutes of the proceedings at all meetings of the committee. He shall draw out a balance sheet every year, shewing the state of the Society's funds, and make out a rate of dividend to each member at any time it is considered requisite to divide any surplus stock.

The *Markers* shall be required to be competent judges of cattle, and each one shall have a certain district allotted to him by the committee, and shall attend to mark any cattle within six days after receiving notice, or be fined 1s.

Marking Cattle.

9.—Each marker will be provided with an Iron for the purpose of marking cattle. The mark must be burned on the horn or the hoof, and the owner shall pay the marker 4*d.* per head for marking. Should the markers go to see any cattle and not think proper to mark them, the Society will pay the 4*d.* If an application be made to mark any cattle which the markers find to be unsound or diseased, or under the value of £10, the matter shall be made known to the President on or before the next club night, and he shall bring it before the committee then present, who will decide whether or not such be accepted, and upon what terms.

If the markers knowingly mark any cattle that are unsound, and the same be proved against them, they shall forfeit the sum of Twenty Shillings. And if any member shall apply to have a diseased one marked, and know the same, shall, on proof being made, forfeit the sum of Twenty Shillings, and not be entitled to any benefit for the loss of such cattle.

Should any marks wear out, the owner shall be under the obligation of paying to have the marks replaced, in order to avoid fraud and error; or in default thereof will not be entitled to benefit, if on inspection no marks are discernible.

Benefits, etc.

10.—No member of this Society shall have any relief from it if he or she wilfully neglect their cattle in not calling in a Veterinary Surgeon, or person skilled in the diseases of cattle, who shall be paid by the owner, at his expense; but if the cattle grow worse or become dangerously ill, the member shall give notice to the Steward, who shall go within eighteen hours after notice given, to view any cattle taken ill, or forfeit the sum of two shillings and sixpence each; and, on viewing the cattle, if the Steward shall think any other V. S. more skilful than the one employed, they shall have power to order the owner of the cattle to employ the person they recommend; and, if the owner do not as ordered, it shall be deemed neglect, and he shall not be entitled to any sum from the fund; but if care be taken, and such cow happen to die, the owner thereof shall receive from the Treasurer of this Society the sum of Ten Pounds, on producing a certificate signed by the Steward or Stewards; but the benefit arising from the sale of the diseased beast shall, in every instance, be paid into the box, and the Stewards shall have the sole right of selling such beast, and shall see such beast is slaughtered.

11.—If any member of this Society keep a cow so marked to old age, to be under the value of five pounds, and the same happen to die, the Stewards shall allow the full amount, or only the real value, as they consider fair and just.

12.—No complaint shall be heard unless proved at the time such neglect shall happen, or, at least, before any claim can be made upon the Society for loss of cattle; and any member making any complaint afterwards shall forfeit the sum of two shillings and sixpence.

13.—When the fund shall fall short of paying the loss for cattle belonging to the members, it shall be lawful for the Stewards to make a rate to pay the money wanting, providing it does not exceed one shilling per cow, which rate shall be paid at the succeeding meeting, over and above the usual subscription, and every member refusing shall be excluded.

14.—Any member purchasing cattle of any description of any member of this Society, already marked, shall not be entitled to any relief from this Society, except the same be viewed, approved, and marked again by the Markers of this Society.

15.—Any member parting with any cattle that are marked according to the Rules aforesaid and buying fresh ones in their stead, if sound, and they pay the Markers for marking them, they shall be entered instead of those parted with, and shall be entitled to benefit, immediately.

16.—If any member be found to have in possession more marked cattle of any description than has been paid for the preceding month or months, by the Markers of this Society, he or she so offending to be entitled to no relief from the Society, and to be excluded.

17.—If the Treasurer has not sufficient funds in hand to pay a member's claim for loss, the same shall be paid at the next monthly night.

18.—No member shall be allowed any benefit from this Society who has cows in this Club belonging to other people entered in his name.

19.—No member shall be allowed to dress' his or her cattle with mercury or mercury water, or any injurious in-gredient whatever: for if any cattle sustain any injury by such application, he or she so offending shall receive no benefit for any cattle so dressed.

20.—A Committee of eleven members shall be chosen annually, by a majority of the members then present, whereof the President shall be one, and all grievances or differences that may arise between members of this Society at any of their monthly or quarterly nights, shall be decided by a majority of the committee then present; and if any member shall upbraid any of the officers or other members, either publicly or privately, concerning club affairs, he shall, upon proof thereof, forfeit one shilling, or suffer such other punishment as the committee shall think fit. Members refusing to serve on the committee when elected shall be fined 6*d.*

21.—Any member who shall curse, swear, or otherwise behave indecently during business hours shall be fined 6*d.*

22.—A Calf Club shall be held in connection with this

Society, and every person entering calves shall pay one shilling entrance, and sixpence per month contribution for each calf. .

23.—No calf shall be entered before the 1st January, and shall not be entitled to any benefit from the Society until after the expiration of fourteen days from the time of entry, when they shall, in case of the loss of such calf or calves, on producing a certificate signed by the Stewards, receive from the Treasurer, if before the 1st of May, the sum of Four Pounds; and on and after the 1st of May, the sum of Six Pounds; providing that Rule 10, as far as regards calling in a V.S., giving notice to the Stewards, also selling of the beast in case of death, shall have been observed, otherwise no claim can be entertained.

24.—No calf shall be marked that is not considered by the Markers to be of the value given in case of a loss, as in the preceding Rule, viz. :—from the 1st January to 1st May, £4; and from 1st May to 30th December, £6.

25.—Calves may be marked at any time into the Calf Club, after the 1st January, providing they be of the value laid down in preceding Rule : but no calf or heifer shall, under any circumstances whatever, be marked into the Cow Club until they are two years old, *i.e.*, not before the 1st of January, the usual time of marking such calves or heifers into the parent Society.

26.—Any calf that has been marked into the Calf Club, and whose contributions are all paid up to the 31st December, shall, on the 1st January, be transferred into the Cow Club, providing such calf or heifer be of the value of ten pounds, and shall be entitled to benefit immediately they are transferred; but should any calf or heifer so marked, and whose contributions are all paid as specified above, but which is thought by the Markers not to be of the value of ten pounds, or is in any way diseased, so as not to be in a proper state to be marked, in order to transfer them into the Cow Club, the case shall be immediately brought before the committee, and it shall be deemed lawful for them to accept such heifer or heifers at such a valuation as they may agree upon; and in case of the loss of such heifer or heifers, the owner thereof shall receive no more from the Society than the amount at which they were valued by the committee at the time they were accepted by them; but should such heifers afterwards attain to the value of ten pounds, either by growth or recovery from disease, the same shall be inspected by the Markers, and if approved of by them, they shall be remarked, and at once be admitted to full benefit, and receive in case of loss the sum Ten Pounds, according to Rule 10.

27.—No Marker or any member of this Society shall at any time mark their own cattle, and any marker or other member so doing, contrary to this Rule, shall not receive any benefit from this Society, in case of the loss of such cattle so marked.

28.—The following salaries shall be paid annually, to the officers of this Society, for their services, viz. :—

Secretary	.	.	.£4	0	0	Markers (each)	.	.£0	10	0
President	.	.	0	10	0	Chapel Cleaner (for cleaning club				
Stewards (each)	.	.	1	0	0	room & lighting fires)	0	5	0	

Such salaries to be paid on monthly club nights in May of each year.

29.—The committee shall have power to alter at any time, any of the foregoing Rules, if requisite; also to make and adopt any new ones they may deem necessary.

30.—Every member of this Society shall stand to, and abide by these Rules; and any member opposing the same, contrary to law, shall be fined the sum of five shillings, or suffer such other punishment as the committee shall think fit, which fines shall go to the box.

MY SMALL DAIRY FARM.*

By F. IMPEY, of BIRMINGHAM.

ALTHOUGH following other occupations, I have for some years past had a small dairy farm of twenty acres in the first place, and five-and-twenty acres at present, from which it has been my object to obtain the largest return possible. Beginning with the plans I found adopted by the farmers around me, I quickly discovered my farming must result in more loss than I could afford if continued on these lines. I therefore made a point of studying the methods of the small farmers of the Continent of Europe, especially those of Belgium and Northern France, with the result that in the first year I increased the balance available for rent from my twenty acres from £30, which it had been under the old-fashioned English plan, to £100. Labouring small farmers prevail in every country in Europe, and in some countries they are the main-stay and backbone of the State. English agricultural

* The following remarks were originally addressed to an audience of labourers farming allotments of one or two acres at Granborough in Buckinghamshire.

labourers are happily beginning to demand
that legal arrangements which shut them out
from a fair chance in life shall be abolished,
and I have no doubt that we can do in
England on our land all that can be done
abroad. Agricultural labourers are already
showing by their allotments, on which they
grow double the average crops of the country
generally, what can be done by hard work, and
skill. The chief thing to be borne in mind,
both in growing cereal crops and in dairy
farming on a small scale, is that the plans
which succeed for a large farm will not succeed
for a small one. We must copy the foreigner
until we can improve upon him. Mr. Jenkins,
the Royal Commissioner on Agriculture, who
visited Belgium and other countries for the
purposes of the Commission, mentions the case
of a farm of ten acres, which supported a
mother and her two sons. They worked very
hard and lived hard, but in an interval of ten
to fifteen years, which passed between his first
and second visits, there had been £200 saved
and invested on the farm. Mr. Jenkins says
the lot of an English agricultural labourer
would be preferable, but all I ask is that the
Englishman shall be allowed the chance of
hiring or buying ten acres for himself, and let
us see whether he will not do it. Again, the
reports furnished by the Belgian landowners
and other persons of influence show in the
great majority of cases that the holdings of

the small farmers of from five to ten acres are
the best cultivated, and that the occupiers had
succeeded the best in struggling against the
bad seasons of the last few years. Mr. Thornton, who wrote an interesting book on "Small
Farming," gives the following particulars of
the stock kept on three Belgian small farms
which he himself visited :—On a ten-acre farm
were four cows, two calves, horse, and two
pigs. On a thirty-eight-acre farm were a bull,
six cows, two heifers, horse, and seventy-five
sheep. On a farm of thirty-two acres were
eight cows, six bullocks, calf, and four pigs,
and the tenant spent every year £200 on
manure. But we need not go to Belgium to
find these facts. In our own island of Jersey we
find one cow is kept to one and three-quarter
acres of ground, as compared with one cow to
five acres in England. And although it is true
that the Jersey cow is much smaller than her
English sister, she probably produces as great
a value in return every year as the larger
English animal, and my experience is that you
can keep three Jersey cows to two average
English cows. So that taking this into account
the balance is still a long way against the
English method, and, as Mr. Thornton, the
great shorthorn authority, has asked, why
should this continue? Canon Bagot, a genial
and clever Irish clergyman, who was sent a few
years ago to report on the dairying of the
Continent, has reported that in every instance

he found the best butter being produced
from land under tillage, and at the present
moment the best Danish butter, which com-
mands as high a price as English, is being
produced from stall-fed cows on arable farms.
My own experience is shortly as follows:—
When I began I kept on my twenty acres of
land four cows and one horse, besides pigs
and poultry. I made hay of ten acres and
grazed the remaining ten. The result of this
was a balance in hand, after paying wages and
all expenses, except rent, of £30. Dissatisfied
with this, and reading what the results of stall-
feeding were, the next season I resolved to
keep my cows in and cut for them. I found
I could then keep seven cows instead of four,
in addition to the horse, pigs, and poultry,
and at present on twenty-five acres of grass
ground I keep an average stock all the year
round of ten cows, two horses, donkey, four
calves, and six pigs, besides poultry. The
following is a calendar of operations:—During
January and February the cows are out by
days, weather permitting, being fed when in-
doors on chaffed hay, with which about 14lbs.
a day of swedes or mangel is mixed, and each
cow is allowed 2lbs. per day of decorticated
cotton cake, which costs £7 or £8 a ton.
New milch cows receive in addition 3 or
4 lbs. of maize or barley meal per day.
During March, April, and most of May we
continue feeding on the dry food, and are very

thankful when, during May, we get our first cutting of prickly comfrey, which is chaffed up in the hay and straw, and eagerly eaten by the cows. By June we are having the first cuts of fresh grass from sheltered corners, and during that month and July we cut as much grass as the cows care to eat every day for them, taking care to get a double quantity on Saturday, to last until Monday morning. By the end of July we turn the cows out by days on to the fresh young aftermath which has grown after the grass which has been already cut for them, tying them up at night and feeding them on cut grass as usual. Continuing to mow the second crop of grass through August and September, we finally leave off cutting grass for the season by October, when the cows begin again to need a little dry hay. They receive 2 lbs. of decorticated cotton cake every day, summer and winter, each cow consuming during the year 7 or 8 cwt., and as the manurial value of this cake is especially high, being valued as worth at least £4 per ton, there should be a valuable manure heap in the yard at the end of the season. One of the great advantages of stall-feeding will be found to be that the manure, instead of being wastefully dropped about the fields, is all collected in one spot, whence it can be drawn to wherever required, and the liquid manure, being conveyed to a tank furnished with an inexpensive pump, is by it pumped into an old wine barrel, mounted on two wheels (which I was glad to think I was thus putting to a better use than it had ever been before), and conveyed to the fields, where its effects on the turf are most striking. By these means the chief part of the constituents of the soil which the cow has removed by being fed upon it are restored to it again and its condition maintained. I find I can feed eleven cows, two horses, and a donkey on two cart-loads of grass per day, averaging a quantity of about 35 or 40 cwt. I find that on an average I feed a cow for a day on about 40 yards square of grass when it is at its heaviest growth, and that in practice a half-acre per cow is ample for feeding a cow for ten or twelve weeks until the lattermath grass is ready for them to go and graze on. As I before said, when the cows were grazing I used to find that four of them trampled down and got very tired of ten acres by the time hay was over, which with us is very late, not beginning until July, their requirements being then, when turned out to graze, two and a half acres each, instead of one and a half acre as at present when stall-fed. The cows are usually taken to water to a pool about a quarter of a mile away, to stretch their legs, and are every morning brushed down with an old bass broom, which serves for curry-comb, and which they greatly enjoy. Their health is always excellent, and they never show any unwillingness to return to their shed. The contrary is often the case when, as so often happens, insects are troublesome. The grass, being fresh each day up to the last, must necessarily be more appetising, and maintain a more regular flow of milk than is the case when the cow, being out on the same pasture day after day, becomes tired of her food, whereas when tied up its freshness tempts her to eat, and so makes milk or flesh or both. In November we give a few cabbages, four or five per day for each cow, which have grown in the garden, and chaff up with the hay, and we buy a few roots for the spring months during which they are tied up; all of these things, from their costliness to buy, being used rather as luxuries than as main articles of diet. You will want to know what buildings are required for the stock I have described. My cow-house is built with brick pillars about eight feet apart, and at the corners. The ends of the buildings are brick also. Between the pillars I fill up the space with timbers. The cows' standing-places are timbers, and the passage up the shed is brick. The roof is galvanised iron, lined underneath with thin boards for warmth. The cost of my cow-house for fourteen cows, with five roofed-in pig-sties, grass-house for the grass in summer, or hay in winter, etc., together with fence to yard, and liquid manure tank to hold 500 gallons, constructed on the plan described, was about £130. The buildings are neat-looking, and will last twenty years, and probably much more. Of course they could be built in brick, and tiled or slated at a much greater cost— probably twice or three times as much. My object was to do everything economically yet lastingly. In practice I find the outlay for litter when the cows lie on timber very trifling —not more than 5s. per cow per year. In summer there is refuse grass, and in winter refuse chaff or hay, which, together with a little saw-dust, costing 6d. per week for six or eight cows, looks very tidy and comfortable. It will be seen the outlay for buildings of this description was about £10 per cow, and an equal amount might be allowed for a stable, straw and chaff house, and calf-pen.

Well, and now what does a cow yield in return for all the trouble taken? One pint of new milk at twopence equals in nutritious value five ounces of beefsteak at fourpence. There can be no doubt then that for a poor man milk is a cheap food for his children, and if combined with some oatmeal porridge, a pint of skim milk for one penny and a halfpenny worth of oatmeal will make a meal on which, as experience has shown, the most vigorous health can be kept and the hardest work done. The wholesale price of new milk for delivery to town milkmen is usually about eightpence per gallon, taking the year through, and in practice it is found that if the milk is kept at home and made into butter, and the skim milk kept for pigs, the return per gallon is about the same. I find from experiment that my pigs give a return of two-pence per gallon for skim milk, so that if you can sell your milk retail at threepence, or, better still, fourpence per quart, it is much the most profitable way of disposing of it. One great advantage of skim milk is its ability to rear calves, which seem to do better on their skim milk than anything else. A calf should be reared for every two cows kept, retaining those whose mothers are the most valuable. There is no such sure way as this of always having a profitable herd of cows. At the present time I have just had the calves I reared last year and those I reared this valued by a practical farmer, who said they were worth (the seven) £70. I have paid £7 10s. for the keep of the three oldest away from home for three months last summer, and I estimate they each had about 200 gallons of skim milk in being reared. Besides this, they have one pound of cake per day after being weaned at four months old, and this will bring their cost out of pocket to £32 altogether, leaving a balance of £38, or £19 per year for whatever they consume in roaming over the pastures in the autumn and winter, or a return of nearly £1 per acre. Another important point to remember is the way in which cows vary in the quality of the milk they give. The following table gives the percentage of cream

15

on the milk of each of eleven cows last
July :—

1	....	25 per cent.
2	....	15 ,,
3	....	8 ,,
4	....	16 ,,
5	....	14 ,,
6	....	14 ,,
7	....	9 ,,
8	....	10 ,,
9	....	16 ,,
10	....	12 ,,
11	....	8 ,,

I found that taking the Jerseys by themselves their milk averaged eighteen per cent.
of cream, half-bred Jerseys fifteen, and shorthorn cows ten and half per cent. In considering these figures, it should be remembered, the
quantity as well as quality of the milk must be
taken into account, and so I worked out the
calculation further. Taking the cream as
worth sixpence per pint and the milk twopence
per gallon, I found that the Jersey cows gave
an average return of two shillings and twopence each per day, the half-bred Jerseys two
shillings and fivepence, and the short-horns
two shillings and a halfpenny. My own
choice for a breed for a dairy herd would be
half-bred Jerseys, combining, as they generally
do, quality and quantity, and an ability to fatten ;
but it is plain that inasmuch as three Jerseys
could be kept on the food of two shorthorns,
there is something to be said in favour of the

little Island cattle. My best Jersey gives me
butter at the rate of 12 to 13 lbs. per week, and
is the most profitable cow that I have. These
figures will illustrate the value of exact knowledge as to what each of our cows is doing,
and we find the foreign small farmer fully alive
to this, Mr. Jenkins giving in his report on
Belgian farming copies of most carefully
arranged returns as to quantity and quality
of each cow's milk which he found in common
use there. If milk is used for butter, let it be
kept scrupulously clean and sweet in every
particular. When churning, especially in
winter, it will be found to be of great assistance to have a thermometer. Before putting
our cream into the churn we raise its tempera-
ture to 65°, and the butter speedily comes—
never taking over half-an-hour. It is best to
stop churning and wash the butter when the
particles are no larger than a pin's head ; the
flavour and keeping-power of butter treated in
this way are greatly superior to that which
has been knocked about in the churn into
lumps and then washed and made up. On
large farms abroad a Laval's separator is
frequently kept, by means of which the cream
is taken from the milk as soon as milked, and
the butter is then churned and made up in
a butter worker ; but my object has been to
describe the most profitable methods yet
known of working a small dairy farm, and I am
always hoping to improve my knowledge in these

matters. I cannot help thinking that on the accomplishment of the political changes which everybody believes to be imminent, the labouring classes of England, both in country and town, will require to know why laws should exist which practically debar them from the enjoyments and profits of small farming. They will even perhaps insist on a Royal Commission in some time to come to report on the best methods of small farming, and demand that dairy schools such as those of Denmark and Germany should be established in this country under the control of the State, where the man who desires to learn the best methods of small farming can do so at cost price. Surely these things are at least of equal importance with the subjects which now engross public notice ; if attended to they will be found to produce their blessed result of an increasing agricultural population—prosperous on the fruits of their own exertions, and furnishing the larger farmer with a supply of skilful and steady servants.

FREDERICK IMPEY.

OPINIONS OF THE PRESS

ON THE

SOCIAL SCIENCE SERIES.

———≫╪≪———

"'The Principles of State Interference' is another of Messrs. Swan Sonnen-schein's Series of Handbooks on Scientific Social Subjects. It would be fitting to close our remarks on this little work with a word of com-mendation of the publishers of so many useful volumes by eminent writers on questions of pressing interest to a large number of the com-munity. We have now received and read a good number of the handbooks which Messrs. Swan Sonnenschein have published in this series, and can speak in the highest terms of them. They are written by men of con-siderable knowledge of the subjects they have undertaken to discuss; they are concise; they give a fair estimate of the progress which recent dis-cussion has added towards the solution of the pressing social questions of to-day, are well up to date, and are published at a price within the resources of the public to which they are likely to be of the most use."—*Westminster Review*, July, 1891.

" The excellent 'Social Science Series,' which is published at as low a price as to place it within everybody's reach."—*Review of Reviews*.

" A most useful series. . . . This impartial series welcomes both just writers and unjust."—*Manchester Guardian*.

" Concise in treatment, lucid in style and moderate in price, these books can hardly fail to do much towards spreading sound views on economic and social questions."—*Review of the Churches*.

" Convenient, well-printed, and moderately-priced volumes."—*Reynold's News-paper*.

" There is a certain impartiality about the attractive and well-printed volumes which form the series to which the works noticed in this article belong. There is no editor and no common design beyond a desire to redress those errors and irregularities of society which all the writers, though they may agree in little else, concur in acknowledging and deploring. The system adopted appears to be to select men known to have a claim to speak with more or less authority upon the shortcomings of civilisation, and to allow each to propound the views which commend themselves most strongly to his mind, without reference to the possible flat contradiction which may be forthcoming at the hands of the next contributor."—*Literary World*.

"'The Social Science Series' aims at the illustration of all sides of social and economic truth and error."—*Scotsman*.

———————

SWAN SONNENSCHEIN & CO., LONDON.

SOCIAL SCIENCE SERIES.

SCARLET CLOTH, EACH 2s. 6d.

SOCIAL SCIENCE SERIES—(*Continued*).

39. **The London Programme.** SIDNEY WEBB, LL.B.
" Brimful of excellent ideas."—*Anti-Jacobin.*
40. **The Modern State.** PAUL LEROY BEAULIEU.
"A most interesting book; well worth a place in the library of every social inquirer."—*N. B. Economist.*
41. **The Condition of Labour.** HENRY GEORGE.
" Written with striking ability, and sure to attract attention."—*Newcastle Chronicle.*
42. **The Revolutionary Spirit preceding the French Revolution.**
FELIX ROCQUAIN. With a Preface by Professor HUXLEY.
" The student of the French Revolution will find in it an excellent introduction to the study of that catastrophe."—*Scotsman.*
43. **The Student's Marx.** EDWARD AVELING, D.Sc.
"One of the most practically useful of any in the Series."—*Glasgow Herald.*
44. **A Short History of Parliament.** B. C. SKOTTOWE, M.A. (Oxon.).
" Deals very carefully and completely with this side of constitutional history."—*Spectator.*
45. **Poverty: Its Genesis and F⸗ 'us.** J. G. GODARD.
" He states the problems with great force and clearness."—*N. B. Economist.*
46. **The Trade Policy of Imperial Federation.** MAURICE H. HERVEY.
"An interesting contribution to the discussion."—*Publishers' Circular.*
47. **The Dawn of Radicalism.** J. BOWLES DALY, LL.D.
" Forms an admirable picture of an epoch more pregnant, perhaps, with political instruction than any other in the world's history."—*Daily Telegraph.*
48. **The Destitute Alien in Great Britain.** ARNOLD WHITE; MONTAGUE CRACKAN-
THORPE, Q.C.; W. A. M'ARTHUR, M.P.; W. H. WILKINS, &c.
" Much valuable information concerning a burning question of the day."—*Times.*
49. **Illegitimacy and the Influence of Seasons on Conduct.**
ALBERT LEFFINGWELL, M.D.
" We have not often seen a work based on statistics which is more continuously interesting."—*Westminster Review.*
50. **Commercial Crises of the Nineteenth Century.** H. M. HYNDMAN.
" One of the best and most permanently useful volumes of the Series."—*Literary Opinion.*
51. **The State and Pensions in Old Age.** J. A. SPENDER and ARTHUR ACLAND, M.P.
"A careful and cautious examination of the question."—*Times.*
52. **The Fallacy of Saving.** JOHN M. ROBERTSON.
" A plea for the reorganisation of our social and industrial system."—*Speaker.*
53. **The Irish Peasant.** ANON.
"A real contribution to the Irish Problem by a close, patient and dispassionate investigator."—*Daily Chronicle.*
54. **The Effects of Machinery on Wages.** Prof. J. S. NICHOLSON, D.Sc.
"Ably reasoned, clearly stated, impartially written."—*Literary World.*
55. **The Social Horizon.** ANON.
" A really admirable little book, bright, clear, and unconventional."—*Daily Chronicle.*
56. **Socialism, Utopian and Scientific.** FREDERICK ENGELS.
" The body of the book is still fresh and striking."—*Daily Chronicle.*
57. **Land Nationalisation.** A. R. WALLACE.
" The most instructive and convincing of the popular works on the subject."—*National Reformer.*
58. **The Ethic of Usury and Interest.** Rev. W. BLISSARD.
" The work is marked by genuine ability."—*North British Agriculturalist.*
59. **The Emancipation of Women.** ADELE CREPAZ.
" By far the most comprehensive, luminous, and penetrating work on this question that I have yet met with."—*Extract from Mr.* GLADSTONE'S *Preface.*
60. **The Eight Hours' Question.** JOHN M. ROBERTSON.

DOUBLE VOLUMES, Each 3s. 6d.

1. **Life of Robert Owen.** LLOYD JONES.
2. **The Impossibility of Social Democracy:** a Second Part of " The Quintessence of Socialism ". Dr. A. SCHÄFFLE.
3. **The Condition of the Working Class in England in 1844.** FREDERICK ENGELS.
4. **The Principles of Social Economy.** YVES GUYOT.

www.ingramcontent.com/pod-product-compliance
Lightning Source LLC
Chambersburg PA
CBHW021659210326
41599CB00013B/1467